BARRON'S
THE TRUSTED NAME IN TEST PREP

AP®
Physics 2
PREMIUM 4TH EDITION

Kenneth Rideout, M.S., and
Jonathan Wolf, M.A., Ed.M.

Published by Kaplan North America, LLC, d/b/a Barron's Educational Series
1515 West Cypress Creek Road
Fort Lauderdale, Florida 33309
www.barronseduc.com

ISBN: 978-1-5062-9201-4

10 9 8 7 6 5 4 3 2 1

Kaplan North America, LLC, d/b/a Barron's Educational Series print books are available at special quantity discounts to use for sales promotions, employee premiums, or educational purposes. For more information or to purchase books, please call the Simon & Schuster special sales department at 866-506-1949.

About the Authors

Ken Rideout has a B.S. in Honors Physics from Purdue University and an M.S. in Physics from Carnegie Mellon University. He has been teaching high school physics for more than twenty years in the Boston area and is currently teaching at Wayland High School in Wayland, Massachusetts.

Jonathan Wolf is an adjunct professor of physics at Fairleigh Dickinson University in Madison, New Jersey. He has been teaching physics at both the secondary school and college levels for more than thirty-five years. He has published more than forty professional papers in the fields of astronomy, physics, and physics education and served for more than ten years as assistant editor for *The Science Teachers Bulletin* published by the Science Teachers Association of New York State (STANYS).

Table of Contents

DIAGNOSTIC TEST

REVIEW AND PRACTICE

PRACTICE TESTS

APPENDICES

Visit Barron's Online Learning Hub for more full-length practice tests.

How to Use This Book

This book provides comprehensive review and extensive practice for the latest AP Physics 2 course and exam.

About the Exam

Start with the Introduction, which outlines the exam format. Review the advice for answering all question types, familiarize yourself with the mathematical relationships and concepts you should know for test day, and learn how to develop a study plan that works for you.

Diagnostic Test

Next, take the short diagnostic test to determine which topics you know well and which ones you may want to brush up on. Complete all of the questions, and then check the answer explanations, especially for any questions you may have missed. If you had difficulty answering any of the multiple-choice questions, refer to the brackets at the end of each answer explanation to see which topic each question covers and the corresponding review chapter in this book.

Review and Practice

Study all seven review chapters, which are divided into the following units of the latest AP Physics 2 course and exam: Thermodynamics; Electric Force, Field, and Potential; Electric Circuits; Magnetism and Electromagnetism; Waves; Geometric Optics; and Modern Physics. Each chapter includes a review of all frequently tested topics, clear figures that illustrate key concepts, sample problems with solutions, helpful tips and reminders, chapter summaries, and end-of-chapter practice exercises with detailed answer explanations.

Practice Tests

There are two full-length practice tests toward the end of the book that mirror the actual exam in format, content, and level of difficulty. Each test is followed by detailed answers and explanations for all questions as well as a test analysis sheet for gauging how well you did.

Appendices

Before completing your review, be sure to consult the Appendices at the end of the book for a table of important information to know for test day, key formulas you should be familiar with, and a glossary of common physics terms and their definitions.

Online Practice

There are also two additional full-length practice tests online. You may take these tests in practice (untimed) mode or in timed mode. All questions are answered and explained.

For Students

In this review book, you will find all the material needed to review and prepare for the AP Physics 2 exam, a second-year precalculus course. This book (and the test itself) assumes you have familiarity with the material in the AP Physics 1 curriculum. AP Physics 2 should be seen not only as additional topics in physics but also as a continuation of AP Physics 1 topics.

Preparing for an AP exam takes time and planning. In fact, your preparation should begin in August or September, when you start the class. If you are using this review book during the year, the content review chapters may parallel what you are covering in class. In that case, study all of the topics within each chapter and attempt every practice exercise, as the questions within these sets vary in style and level of difficulty and are intended to test your level of understanding of the review material. If you are using this review book a few weeks prior to the exam and your study time is limited, your strategy needs to change. Instead, review the end-of-chapter summaries and take both full-length practice exams, which test your knowledge of all the various content areas of AP Physics 2. Be sure to read through all the tips and sidebars throughout the book for helpful advice and reminders for test day, and above all, best of luck on your AP Physics 2 exam!

For Teachers

This book is fully aligned with the redesigned curriculum, units, and exam format outlined in the latest *AP Physics 2 Course and Exam Description*. You can use this book as a resource in the classroom, or you can assign chapters as supplemental reading or practice questions as homework or test material.

BARRON'S ESSENTIAL 5

As you review the content in this book to work toward earning that **5** on your AP PHYSICS 2 exam, here are five things that you **MUST** know above everything else:

1 **Review the key concepts of the AP Physics 1 curriculum** (e.g., review the Essential 5 in Barron's *AP Physics 1 Premium*). AP Physics 2 is designed as a second-year course, and the key ideas of AP Physics 1 provide the essential knowledge base for the material for this exam.

2 **Understand fields:**

- Be able to visualize, draw, and interpret electric and magnetic field lines.
- Be able to make analogies between electric and gravitational fields.
- Understand the difference between the field strengths and the actual force exerted on an object in that field.
- Understand the differences between field potentials and the actual potential energy present when an object is in the field.

3 **Know how and when to use the various models for waves:**

- Model light and sound as a wave for interactions such as diffraction, Doppler shift, and interference and when describing it in terms of amplitude and wavelength.
- Model light as a series of wavefront rays for problems in optics (reflection, refraction, mirrors, and lenses).
- Model light as individual particles (photons) in atomic and modern physics.

Understand and trace the distribution of voltage and flow of current in simple and complex circuits:

- Although detailed questions about the exponential nature of charging and discharging capacitors will not be asked, you must understand their behavior and purpose when fully charged or fully discharged in circuits.

Understand the energy conservation roots of thermodynamics and Kirchhoff's Rules:

- Understand the specialized vocabulary of thermodynamics and be able to connect all operations to energy transfers and transformations.
- Relate Kirchhoff's Loop Rule to the conservation of energy when analyzing circuits.

Introduction

The College Board currently offers four AP Physics exams. This book is for students preparing for the AP Physics 2 exam, which corresponds to a second-year algebra-based college course. The other exams are AP Physics 1 (a first-year algebra-based college course), AP Physics C: Mechanics (calculus-based), and AP Physics C: Electricity and Magnetism (also calculus-based). Note that the old AP Physics B is a retired test and is no longer offered by the College Board. Both the AP Physics 1 and AP Physics 2 exams focus on conceptual underpinnings and basic scientific reasoning along with the traditional problem-solving aspects of physics. In addition, both exams have questions that require experiential lab understanding. Although there are some calculation-oriented questions, these two tests are designed explicitly to not be "plug and chug" questions. If you do not thoroughly understand the physics concepts behind the equations, you will find yourself at a disadvantage.

The AP Physics 1 exam focuses on kinematics, dynamics, the three conservation laws, oscillations, and fluids. The AP Physics 2 exam assumes you have already covered these topics and understand this material. If you have not studied these topics, it is highly recommended that you review them (with the Barron's *AP Physics 1 Premium* book, for example) before using this book. The AP Physics 2 exam not only assumes you have this background information, but it also continues the study of physics with thermodynamics, electric charge and force, magnetism, electromagnetism, introductory circuits, waves, optics, and modern physics.

Conventions of AP Physics 2:

- The frame of reference of any problem is assumed to be inertial.
- Positive work is defined as work done on a system.
- The direction of current is conventional current (the direction positive charges would drift).
- All batteries and meters are ideal (unless otherwise stated).
- Edge effects are assumed for the electric field of a parallel plate capacitor.
- For any isolated electrically charged object, the electric potential is defined as zero out at infinity.

Exam Format

You will have three hours to complete the AP Physics 2 exam, which consists of the two sections outlined in Table A.

Table A AP Physics 2 Exam Format

	Section I	Section II
Question Type	Multiple-Choice	Free-Response
Number of Questions	40	4
Time	80 minutes	100 minutes
% of Overall Score	50%	50%

Formulas are provided for your use during the test (see the Appendices of this book). However, it is important that you not only know what the provided equations mean but also are able to quickly determine in what situations they can and should be used. Even if a question is conceptual, having a corresponding equation in mind can guide your thinking. The equation sheet provides a solid foundation. A well-prepared student will be able to find and understand every equation on the equation sheet.

Although you may not need to use your calculator a lot during the exam, one is allowed throughout. (Check the College Board website for an approved list of calculators. Generally, though, all scientific and graphing calculators are allowed.) Make sure your calculators are fully charged and that you have extra batteries for your calculator during the exam. A ruler is also permitted. Its usefulness is likely limited, though, to drawing straight lines, if needed, during the free-response section.

Section I: Multiple-Choice

Without a doubt, multiple-choice questions can be tricky. The AP Physics 2 exam asks multiple-choice questions that can range from a simple recall of information to questions about units, graphs, proportional relationships, formula manipulations, and simple calculations (without a calculator). The questions cover all areas of the course, and the exam weighting per unit for Section I is outlined in Table B.

Table B Exam Weighting for Section I of the AP Physics 2 Exam

Unit	Percentage
Thermodynamics	15–18%
Electric Force, Field, and Potential	15–18%
Electric Circuits	15–18%
Magnetism and Electromagnetism	12–15%
Geometric Optics	12–15%
Waves, Sound, and Physical Optics	12–15%
Modern Physics	12–15%

Section II: Free-Response

The AP Physics 2 exam includes four free-response questions, each of which is a unique question type requiring specific skills. The general format of Section II is outlined in Table C, followed by a description of each question type.

Table C Section II of the AP Physics 2 Exam

Question Type	Suggested Time	Point Value
Mathematical Routines	20–25 minutes	10 points
Translation Between Representations	25–30 minutes	12 points
Experimental Design and Analysis	25–30 minutes	10 points
Qualitative/Quantitative Translation	15–20 minutes	8 points

The **Mathematical Routines** question will require you to use mathematics to analyze and make predictions about a scenario.

The **Translation Between Representations** question will require you to connect different representations of a scenario and draw graphs that relate quantities within the scenario. Also, you will be asked to make predictions (about different situations or changes to the original scenario) and/or justify your answers or predictions.

The **Experimental Design and Analysis** question will require you to create scientific procedures and analyze data. This question will be broken down into two sections:

1. Design: You will be asked to design a high school lab experiment that is based around a scientific question, varies by only one parameter, and measures how that change affects one characteristic. You will also need to describe how data can be collected to help answer that question.

2. Analysis: You will be given experimental data based on a similar question to that in the Design part, and you will be asked to use that data to create and analyze a graph that will help answer the question.

The **Qualitative/Quantitative Translation** question will focus on a scenario and require that you make connections between the physical laws that govern the scenario and mathematical representations of the scenario. You will need to make and justify claims about the scenario and derive an equation connected to that scenario. Also, you will be asked to make predictions (about different situations or changes to the original scenario) and/or justify your answers or predictions.

Tips for Section I: Multiple-Choice Questions

- **Remember that there is no penalty for wrong answers.** This means that you should try to answer all the questions.
- **Guess logically.** For the AP Physics 2 exam, all multiple-choice questions will have four answer choices and only one correct answer. If you're not sure about the correct answer, instead of randomly guessing, you can improve your chances of getting the correct answer by eliminating at least two answer choices.
- **Look for distractors.** Distractors are choices that may look reasonable but are incorrect. For example, if the question is expecting you to divide to get an answer, the distractor may be an answer obtained by multiplying. Watch out for quadratics (such as centripetal force) or inverse squares (such as Coulomb's law).
- **If a formula is needed, try to use approximations (or simple multiplication and division).** For example, the magnitude of the acceleration due to gravity (g) can be approximated as 10 m/s^2. You can also use estimations or order of magnitude approximations to see if answers make sense.
- **Make connections between questions and topics when possible.** If you cannot recall some information, perhaps another similar question will cue you as to what you need to know. (Note that you may work on only one part of the exam at a time.) When you read the question, try to link it to the overall general topic, such as electricity. Then narrow down the specific area and the associated formula. Finally, you must know which quantities are vectors and which quantities are scalars.
- **Practice both with and without timing.** At the start of your review, you may want to work without a timer on the multiple-choice questions for the diagnostic test and the first practice test. A few days before the exam (see the sample study plan later in this Introduction), you should attempt the second practice test timed.

Tips for Section II: Free-Response Questions

- **Be prepared.** Make sure that you have a working calculator with extra batteries in the days before your exam.
- **Read the entire question carefully and plan your response.** Before solving the problem, make sure you know what the question is asking of you. Write down the general concept being used—for example, conservation of mechanical energy or conservation of energy. Then write down the equations you plan to use. For example, if the problem requires you to use conservation of mechanical energy (potential and kinetic energies), write out those equations:

Initial total mechanical energy = Final total mechanical energy

$$mgh_i + \tfrac{1}{2}mv_i^2 = mgh_f + \tfrac{1}{2}mv_f^2$$

- **Review the equation sheet.** An equation sheet will be provided on exam day. Make sure you know where the formulas and constants can be found on this sheet. One of the first things you may notice is that you are not given every formula you have ever learned. Some teachers may let you use a formula sheet during their classroom exams, and some teachers may require you to memorize formulas. Even if you get to use a formula sheet during a classroom exam, you should memorize derivations and variations of formulas. Furthermore, since you are not given specific formulas for some concepts, you should begin learning how these formulas are derived, starting at the beginning of the school year. For example, you will not be given the specific formulas for projectile motion problems since these are easily derived from the standard kinematics equations. If you begin reviewing a few weeks before the AP exam, you may want to make index cards of formulas to help you to learn them.

- **Show all of your work.** Include all formulas, substitutions, and general concepts used to earn full credit. When you are making substitutions, indicate to the grader what you are doing. For example, if you are calculating the net force on a mass, write as neatly as possible:

$$\sum F = F_{net} = ma = (2\,\text{kg})(4\,\text{m/s}^2) = 8\,\text{N}$$

- **Include all relevant information.** Show the grader that you understand what the question is asking. You may want to make a few sketches or write down your thoughts in an attempt to find the correct solution path. If a written response is requested, make sure you write neatly and answer the question in full sentences.

- **Remember to label all diagrams.** Sometimes the question may refer to a lab experiment typically performed in class or simulated data are given. In that case, you may be asked to make a graph. Make sure the graph is labeled correctly (with axes labeled and units clearly marked), points are plotted as accurately as possible, and best-fit lines or curves are used where appropriate. Do not connect the dots. Always use the best-fit line for calculating slopes. Make sure you include units on all final answers, including slopes. If you are drawing vectors, make sure the arrowheads are clearly visible. For angles, there is some room for variation. If you are asked to draw a free-body diagram, include only actual applied forces; do not include component forces. Centripetal force is not an applied force and should not be included on a free-body diagram.

- **Pay attention to your calculator mode.** Since angles are measured in degrees, be sure your calculator is in the correct mode. If scientific notation is used, make sure you know how to input the numbers into your calculator correctly. Remember, each calculator is different.

- **When you are not sure how to solve a problem, don't panic!** Follow these ten tips:

  1. Make sure you understand the general concepts involved, and jot down those concepts, plus all appropriate equations.
  2. Try to see how this problem may be similar to one you have solved before.
  3. Determine what information is relevant and what information is irrelevant to what is being asked.
  4. Rephrase the question in your mind. Maybe the question is worded in a way that is different from what you are used to.
  5. Draw a sketch of the situation if one is not provided.
  6. Write out what you think is the best way to solve the problem. This sometimes triggers a solution.
  7. Use numbers or estimations if the solution is strictly algebraic manipulation, such as deriving a formula in terms of given quantities or constants.
  8. Relax. Sometimes if you move on to another problem, take a deep breath, close your eyes, and just relax for a moment, the tension and anxiety may go away and allow you to continue.
  9. Do not leave anything out. Unlike on the multiple-choice questions, you need to show all of your work to earn credit.
  10. Understand what you are being asked to do. The AP Physics 2 exam requires you to respond in specific ways to certain key words or "task verbs":
      - **Calculate**—provide numerical and algebraic work leading to the final answer (and don't forget units and significant figures!)
      - **Compare**—elaborate on similarities and/or differences
      - **Derive**—mathematically manipulate a fundamental equation (such as those given on the equation sheet) into the desired form
      - **Describe**—list the relevant characteristics
      - **Determine**—after explaining or calculating, arrive at a conclusion

- ○ **Draw**—create a diagram that shows physical objects and their relationships
- ○ **Estimate**—roughly calculate (to the closest power of 10), indicate greater/less than, or indicate positive/negative values (no need to show work)
- ○ **Indicate**—simply provide information (without explanation)
- ○ **Justify**—provide qualitative (not mathematical) reasons to support your claim
- ○ **Label**—indicate unit, scale, or components in a graph or other representation
- ○ **Plot**—place specific data points onto a scaled grid; do not connect the dots (although trends, especially linear ones, may be superimposed on the graph)
- ○ **Rank**—order by magnitude
- ○ **Sketch**—without numerical scaling or specific data points, draw a representation that captures the key trend in the relationship (curvature, asymptotes, and so on)
- ○ **Verify**—show that the specific condition is met and explain why it applies

Scoring of the AP Physics 2 Exam

The AP Physics 2 exam is scored on a scale from 1 to 5, with 5 being the highest possible score. Table D describes each score.

Table D AP Physics 2 Exam Scores

AP Exam Score	Recommendation
5	Extremely Well Qualified
4	Well Qualified
3	Qualified
2	Possibly Qualified
1	No Recommendation

Scores of 3 or above may earn you college credit. Policies regarding credit for AP exam scores vary widely among schools and may even vary among majors at the same school. Always check with the college or university you plan to attend to find out the latest information.

Each section of this exam is worth 50 percent of your grade to determine your "raw score." Keep in mind that the curve for the exam changes from year to year.

Units

One of the most important things to remember for the AP Physics 2 exam is that most physical quantities have units associated with them. You must memorize units since you may be asked questions about them in the multiple-choice section. In the free-response questions, you must include all units when writing final answers.

A list of standard fundamental (SI) units as well as a list of some derived units are shown in the following two tables. As you work through the chapters in this book, make a note (on index cards, for example) of each unit.

TIP

Make sure you memorize all units. Be sure to include them with all calculations and final answers.

Table E Fundamental SI Units Used in Physics

Quantity	Unit Name	Abbreviation
Length	Meter	m
Mass	Kilogram	kg
Time	Second	s
Electric current	Ampere	A
Temperature	Kelvin	K
Amount of substance	Mole	mol
Luminous intensity	Candela	cd

Table F Some Derived SI Units Used in Physics

Quantity	Unit Name	Abbreviation	Expression in Other SI Units
Area			m^2
Linear velocity			m/s
Linear acceleration			m/s^2
Force	Newton	N	$kg \cdot m/s^2$
Momentum			$kg \cdot m/s$
Impulse			$N \cdot s = kg \cdot m/s$
Angular velocity			rad/s
Angular acceleration			rad/s^2
Torque			$N \cdot m$
Angular momentum			$kg \cdot m^2/s$
Moment of inertia			$kg \cdot m^2$
Spring constant		N/m	kg/s^2
Frequency	Hertz	Hz	s^{-1}
Pressure	Pascal	Pa	$N/m^2 = kg/(m \cdot s^2)$
Work, energy	Joule	J	$N \cdot m = kg \cdot m^2/s^2$
Power	Watt	W	$J/s = kg \cdot m^2/s^3$
Electric charge	Coulomb	C	$A \cdot s$
Electric field		N/C	$kg \cdot m/(A \cdot s^3)$
Electric potential	Volt	V	$J/C = kg \cdot m^2/(A \cdot s^3)$
Resistance	Ohm	Ω	$V/A = kg \cdot m^2/(A^2 \cdot s^3)$
Capacitance	Farad	F	$C/V = A^2 \cdot s^4/(kg \cdot m^2)$
Magnetic flux	Weber	Wb	$V \cdot s = kg \cdot m^2/(A \cdot s^2)$
Magnetic flux density	Tesla	T	$Wb/m^2 = kg/(A \cdot s^2)$
Inductance	Henry	H	$Wb/A = kg \cdot m^2/(A^2 \cdot s^2)$

Mathematical Relationships

Since AP Physics 2 is an algebra-based course, the Appendices review some essential aspects of algebra. In physics, we often discuss how quantities vary using proportional relationships. Four special relationships are commonly used.

- Direct relationship—This is usually represented by the algebraic formula $y = kx$, where k is a constant. This is the equation of a straight line, starting from the origin. An example of this relationship is Newton's second law of motion, $\vec{a} = \dfrac{\vec{F}_{net}}{m}$, which states that the acceleration of a body is directly proportional to the net force applied.
- Inverse relationship—This is usually represented by the algebraic formula $y = \dfrac{k}{x}$. This is the equation of a hyperbola. An example of this relationship can also be seen in Newton's second law of motion. If a constant net force is applied to a body, the mass and acceleration are inversely proportional to each other. Some special relationships, such as gravitation and static electrical forces, are known as inverse square law relationships. The forces are inversely proportional to the square of the distances between the two bodies.
- Squared (quadratic) relationship—This is usually represented by the algebraic formula $y = kx^2$ and is the equation of a parabola starting from the origin. An example of this relationship can be seen in the relationship between the displacement and uniform acceleration of a mass from rest, $\vec{x} = \frac{1}{2}\,\vec{a}t^2$.
- Square root relationship—This is usually represented by the algebraic formula $y = k\sqrt{x}$ and is the equation of a "sideways" parabola. This relationship can be seen in the relationship between the period of a simple pendulum and its length, $T = 2\pi\sqrt{L/g}$.

Graphs, Fits, and the Linearization of Data

Graphs that are linear in nature are much easier to analyze, especially by hand, than graphs of any other nature. Trends, slopes, intercepts, and correlations of experimental results to theoretical predications are readily obtained. For this reason, if you are asked to graph your data, it will almost always be advantageous to linearize it first. Specifically, if the relationship is not linear to start, use a change of variable to make the relationship linear.

For example, if asked to determine the spring constant k of a system based on a collection of elastic potential energies for various extensions of the spring, the relevant equation is

$$U_s = \frac{1}{2}\,kx^2$$

This is a quadratic relationship, not a linear one. Graphing energy versus displacement will therefore produce a parabola shape from which it is difficult to extract the spring constant k. A better choice is to linearize the data before graphing. Before graphing, make the following change of variable:

$$z = x^2$$

So the relationship is now:

$$U_s = \frac{1}{2}\,kz$$

Now when graphed and a line of best fit is applied, the slope of the straight line will be $\frac{1}{2}k$.

How can a line of best fit be generated by hand? If need be, you can use a straightedge and draw one straight line that has as many data points above the line as below it. Once this line is drawn, all subsequent calculations should be based on the slope and intercept of this best-fit line rather than on the original data. The idea here is that the fit of the data is an average of the raw data and is inherently better than any one particular point because the random variations in data have been smoothed out by the fit. When asked to analyze a graph of data, always use the fitted line or curve rather than the individual data points for this same reason.

❯ Sample Problem

Determine k by graphing the following data provided by another student.

U_S (J)	x (cm)
0.058	2.5
0.196	4.6
1.117	11.0
2.081	15.1

✓ Solution

Since the relationship between these variables is quadratic, begin by squaring the given values for x. Also change to the standard SI units of meters as well.

x (cm)	z (m^2)
2.5	0.000625
4.6	0.002116
11.0	0.0121
15.1	0.022801

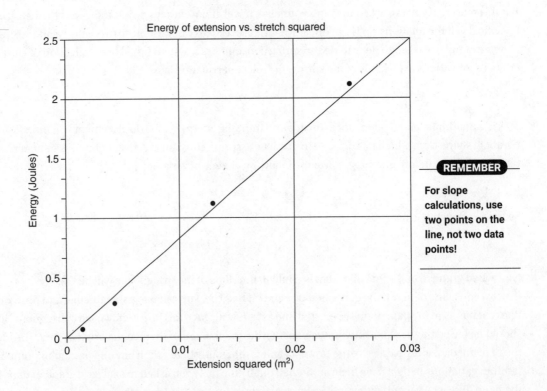

Energy of extension vs. stretch squared

REMEMBER

For slope calculations, use two points on the line, not two data points!

Approximate slope $= \Delta y / \Delta x = 2.5/0.03 = 83$ N/m (since J $=$ N $\cdot$ m)

Note the labels (with units) on the x- and y-axes as well as the title for the graph.

Uncertainty and Percent Error

In addition to respecting the number of significant digits in measured or recorded data, you can also determine the corresponding uncertainty in derived quantities. For example, if the radius of a circle is measured and recorded as 3.5 cm, there are only two significant digits in this number. Therefore, the area of the circle (πr^2) should be truncated from the calculator result to two significant digits. The area is 38 cm^2. The rest of the digits are not significant as they imply a precision in the radius that we do not have.

To take this analysis one step further, a percent error can be associated with a measurement. For example, one could write down the uncertainty in another radius explicitly as $r = 4.5 +/- 0.05$ cm. This implies that the true value is most likely between 4.45 and 4.55 cm. This is a percent error of 1.1%.

$$\frac{0.05}{4.5} \times 100 = 1.1\%$$

$$100 \times \text{Uncertainty/Value} = \text{Percent error}$$

Percent errors are an easier way of comparing the relative precision of different measurements. For example, a measurement of $128 +/- 2$ mm has a percent error of 1.6% and is thus less precise than our original measurement that has an error of 1.1%.

On the AP Physics 2 exam, you may need to assess or calculate uncertainty, either qualitatively or quantitatively depending on the question type. Additionally, you may need to support your answers by properly using significant digits and percent error. However, you will not be expected to propagate errors, calculate standard deviations, or carry out a formal linear regression.

Objects vs. Systems

An **object** is thought of as an isolated mass that is being acted upon by outside forces. Any internal structure of the object is ignored. The object is defined by its properties (e.g., mass, charge). Strictly speaking, for a single object being analyzed, it does not make sense to make use of potential energy, Newton's third law, thermal energy, or the conserved quantities. Instead, when discussing a single object, use the following relationships from mechanics:

$$\vec{F}_{net} = m\vec{a}$$
$$\vec{\tau} = I\vec{\alpha}$$
$$\vec{F}\Delta t = m\Delta\vec{v}$$
$$F\Delta d = \Delta\text{KE}$$

A **system**, in contrast, is made up of objects that may interact. It is within isolated systems that the conserved quantities, canceling Newton's third law forces (internal forces), the heat capacity of the constituent parts, and the potential energy of the relationship between interacting objects are useful concepts. Within systems, conserved quantities become the most useful lens through which to view the situation:

$$(\vec{p}_{net})_i = (\vec{p}_{net})_f$$
$$(\vec{L}_{net})_i = (\vec{L}_{net})_f$$
$$\text{ME}_i - \text{ME}_f$$

Fundamental Particles

A **fundamental particle** is a true object. In other words, if an object is found to have no internal structure (i.e., it cannot be broken into smaller pieces), then it is a fundamental particle. Electrons and photons are fundamental particles, whereas protons and neutrons are not. Electrons and photons can be converted into energy but cannot be broken into smaller pieces. Neutrons and protons, however, can be broken into their constituent parts (quarks and gluons, which are fundamental particles themselves). Einstein's famous $m = E/c^2$ is the course for the majority of mass contained within systems. (The binding energies between the fundamental particles provide the energy.) The search for mass belonging to some fundamental particles themselves (rather than arising from the energy contained within) culminated in the 2013 Nobel Prize in Physics being awarded to two physicists who first proposed a mechanism by which fundamental particles themselves acquire mass (the Higgs field).

Modern theoretical physics (the standard model) is predicated on this notion of a limited number of fundamental particles combining in various ways to give rise to all matter. These fundamental particles influence each other by exchanging other fundamental particles of interaction. Just as all molecules are made up of a limited number of elements in the periodic table, so is everything in the universe made up of a limited number of types of interactions among a limited number of fundamental particles.

Test-Taking Advice and Developing a Study Plan

Preparing for any AP exam takes practice and time. Effective studying involves managing your time so that you efficiently review the material. Do not cram a few days before the exam. Getting a good night's sleep before the exam and having a good breakfast the day of the exam is a better use of your time than "pulling an all-nighter." Working in a study group is a good idea. Using index cards to make your own flashcards of key concepts, units, and formulas can also be helpful.

When you study, try to work in a well-lighted, quiet environment when you are well rested. Studying late at night when you are exhausted is not an effective use of your time. Although some memorization may be necessary, physics is best learned (and studied) by actively solving problems. Remember, if you are using this book during the year, working through the chapter problems as you cover each topic in class, memorizing the units, and familiarizing yourself with the formulas at that time will make your studying easier in the days before the exam.

If you are using this book in the weeks before the exam, make sure you are already familiar with most (if not all) of the units, equations, and topics to be covered. You can either use the chapter review for a quick overview and practice or dive right in to the diagnostic test. You do not need to take the diagnostic test under timed conditions. See how you do, and then review the concepts for those questions that you got wrong. You can use the end-of-chapter questions (some of which may be more difficult than those on the actual AP exam) to test your grasp of specific topics and then work on the remaining practice tests.

Setting up a workable study schedule is also vital to success. Each person's needs are different. The following schedule is just one example of an effective plan.

Table G Sample Study Plan

August/ September–April	As the year progresses, make sure you memorize units and are comfortable with formulas. If you are using this book during the year, complete the end-of-chapter problems as they are covered in class. Make sure you register for the exam, following school procedures, and refer to the College Board's website for details: *www.collegeboard.org*
Four weeks before the exam	Most topics should be covered by now in class. If you are using this book for the first time, begin reviewing concepts and completing the end-of-chapter problems. Begin reviewing units and formulas. Devote at least thirty minutes each day to studying.
Three weeks before the exam	Start working on the diagnostic test. Go back and review topics that you are unsure of or feel that you answered incorrectly.
Two weeks before the exam	Begin working on the practice tests. Continue to review concepts you find more challenging.
One week before the exam	Complete any remaining practice tests under timed conditions. Make sure you are comfortable with the exam format and know what to expect. Review any remaining topics and units.
The day before the exam	Pack up your registration materials, pencils, calculator, and extra batteries. Put them by the door, ready to go. Get a good night's sleep.
The day of the exam	Have a good breakfast. Make sure you take all the items you prepared the night before. Relax!

SUMMARY

- Set up a manageable study schedule well in advance of the exam.
- Make sure you memorize all units and are familiar with the exam format.
- Multiple-choice questions do not have a penalty for wrong answers, so do not skip any. If you are unsure of the answer, try to eliminate as many choices as you can, and then guess!
- Do not leave any question out on the free-response section! Show all of your work. Write down all fundamental concepts, write all equations used, and include units in your final answer.
- Read each question carefully. Write your answers clearly. Use full sentences for your short-answer responses. Clearly label graphs with units, and use best-fit lines or curves.
- An object can be represented as a single mass. When exposed to external forces, it is best modeled as experiencing changes in speed and/or direction.
- A system is a group of objects that, if isolated, is best modeled with conservation laws.
- A fundamental particle has no internal structure.
- Try to complete all of the practice tests. Work on the chapter questions to review concepts as needed.
- Get a good night's sleep before the exam.
- On the day of the exam, bring all registration materials with you, as well as pencils, a calculator, and extra batteries.

Relax and Good Luck!

Diagnostic Test

The following is a short diagnostic test. The purpose of this diagnostic test is for you to identify those conceptual areas most in need of review. The relevant sections of this book (noted in brackets at the end of each answer explanation) should be reviewed thoroughly before attempting one of the full-length practice tests.

Diagnostic Test

Section I: Multiple-Choice

> **DIRECTIONS:** For each question or incomplete statement, select the choice that best answers the question or completes the statement. The brackets at the end of each answer explanation will direct you to a specific chapter for further review on the subject matter covered in that question. You may use a calculator and make use of the formula sheet provided in the Appendices.

1. The Sun's rays strike a black surface that is directly on top of a liquid. The liquid is then observed to swirl as it transfers the heat to the bottom layers of the liquid. The correct sequence of transfers in thermal energy in this story is

   (A) radiative → convective → conductive
   (B) radiative → conductive → convective
   (C) convective → conductive → convective
   (D) convective → conductive → radiative

2. Which of the following is the best explanation of the statement, "A roomful of neon gas has a lower average speed than 1 mole of helium gas at the same temperature and pressure"?

   (A) This is not a correct statement.
   (B) There is more than 1 mole of neon gas in an average room.
   (C) Helium atoms have more degrees of freedom than do neon atoms.
   (D) Helium has a lower mass per particle than neon.

3. Consider the following process for a gas. Which of the following statements is true concerning this process?

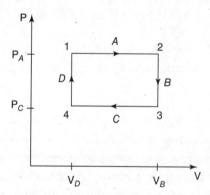

   (A) Path D (from point 4 to point 1) is isobaric.
   (B) No work was done to or by the gas during path A (from point 1 to point 2).
   (C) No net work was done to or by the gas during the entire cycle (from point 1 back to point 1 through A, B, C, D).
   (D) No work was done to or by the gas during path B (from point 2 to point 3).

4. Two identical spheres originally in contact undergo induced charge separation. The two spheres are slowly separated until a gap of 12 cm is separating their surfaces. The net force of attraction between the two oppositely charged large spheres after they have been separated by 12 cm is measured. The entire experiment is then repeated with two smaller but otherwise similar spheres.

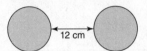 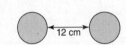

The force between the smaller spheres compared with the force between the larger spheres is best described as

(A) the same since the charges and separations are the same.
(B) smaller since the charges on each small sphere are closer together.
(C) bigger since the centers of the spheres are closer.
(D) the same since the average force felt by any one particular charge is the same in both cases.

5. One large conducting sphere with a net charge of +8 microcoulombs is put into contact with a slightly smaller conducting sphere that initially has a charge of –6 microcoulombs. Which statement best describes the charge distribution after the two spheres are separated?

(A) They are both neutral.
(B) The larger sphere has +8 microcoulombs of charge, while the smaller retains –6 microcoulombs of charge.
(C) Both spheres have +1 microcoulomb of charge.
(D) The larger sphere has slightly more than +1 microcoulomb of charge, while the smaller sphere has slightly less than +1 microcoulomb of charge.

6. Which of the following pictures is an incorrect representation of the electric field near various charged conductors?

(A)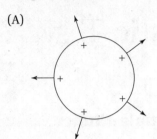

(B)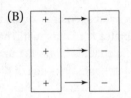

(C)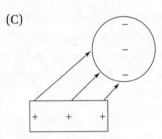

(D)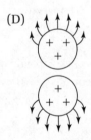

7. A standard RC (resistor-capacitor) circuit takes approximately 2 seconds to charge the capacitor fully. If the resistor is 100 ohms and the voltage across the RC combination is 12 volts, what is the current through the fully charged capacitor?

(A) One must know the capacitance of the capacitor in order to answer this question.
(B) 0.12 A
(C) 0.06 A
(D) 0 A

8. If each of the batteries in the circuit below supplies a voltage V and if each resistor has resistance R, what will be the reading on the meter pictured?

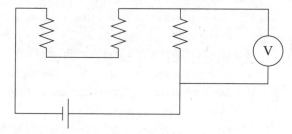

(A) 0

(B) $V/3$

(C) $2V/3$

(D) $2V/3R$

9. The primary difference between a permanent magnet (an object with identifiable north and south poles) and magnetic material that does not have identifiable north and south poles is

(A) electron configuration.

(B) magnetic domain alignment.

(C) magnetic materials have no poles at all.

(D) magnetic materials are lacking one of the two poles.

10. A very long wire carries a steady current. Which of the following statements describes the magnetic field at a distance of R from the wire?

(A) The magnetic field is zero.

(B) The magnetic field loops around the wire and gets weaker proportional to $1/R^2$ as one moves to larger R values.

(C) The magnetic field loops around the wire and gets weaker proportional to $1/R$ as one moves to larger R values.

(D) The magnetic field extends radially outward and gets weaker proportional to $1/R^2$ as one moves to larger R values.

11. Two parallel wires carrying current near each other exert magnetic forces on each other. If the direction and magnitude of the current in both wires is reversed and doubled, what will happen to the magnetic forces?

(A) They will reverse directions and double in magnitude.

(B) They will reverse directions and quadruple in magnitude.

(C) They will remain in the same directions and double in magnitude.

(D) They will remain in the same directions and quadruple in magnitude.

12. A conducting loop of wire is held steady parallel to the ground within a strong magnetic field directed upward. The setup is viewed from above. As the loop of wire is quickly pulled out of a magnetic field, what will the loop experience?

(A) a temporary clockwise current that dies out soon after the loop is out of the field

(B) a temporary counterclockwise current that dies out soon after the loop is out of the field

(C) a temporary clockwise current that dies out as soon as the loop is out of the field

(D) a temporary counterclockwise current that dies out as soon as the loop is out of the field

13. A charge is accelerated by an electric field. While the charge is accelerating, which of the following is occurring?

(A) A static magnetic field is created.

(B) A new, static electric field is created.

(C) Only a varying magnetic field is created.

(D) An electromagnetic wave is created.

14. A light beam is traveling in a vacuum and then strikes the boundary of another optically dense material. If the light comes to the interface at an angle of incidence of 30 degrees on the vacuum side, it will emerge with an angle of refraction

(A) equal to 30 degrees.

(B) less than 30 degrees.

(C) greater than 30 degrees.

(D) The relative indices of refraction must be known in order to find the solution.

15. A beam of light incident at 50 degrees travels from a medium with an index of refraction of 2.5 into one with an index of refraction of 1.1. The refracted angle of light that emerges is

   (A) 20 degrees.
   (B) 50 degrees.
   (C) 90 degrees.
   (D) There is no solution.

16. A radio telescope uses a curved reflecting material to focus radio signals from orbiting satellites or distant objects in space onto a detector. Which of the following would be the biggest problem with designing a radio telescope with a convex reflecting dish?

   (A) The image produced is virtual.
   (B) The image produced is inverted.
   (C) The detection equipment would have to be placed in the line of sight.
   (D) There is no real problem with designing a convex reflecting dish for radio waves.

17. When examining a two-slit diffraction pattern, which of the following would you notice as the slits were brought closer together?

   (A) The interference pattern would contain more maximums. They would be more closely spaced.
   (B) The location of the central interference peak would shift.
   (C) The interference pattern would spread out.
   (D) The interference peaks themselves would each become sharper (narrower).

18. A radioactive isotope undergoes gamma decay. Afterward, the nucleus

   (A) contains fewer nucleons.
   (B) has the same number of nucleons but a different ratio of protons/neutrons.
   (C) has the same mass number but a lower mass.
   (D) undergoes beta decay.

19. When examining the spectrum for a specific element, a bright spectral line with the highest frequency represents

   (A) an electron excited to the atom's highest energy state.
   (B) an electron falling from the highest energy state in the atom to the next to highest state.
   (C) an electron excited from its ground state to the next to lowest energy state.
   (D) an electron falling from a higher energy state to a much lower energy state.

20. Some objects can be made to exhibit an interference pattern with the two-slit experiment, and some objects cannot. What is the difference between those that can and those that cannot?

   (A) Objects are either fundamentally a wave or fundamentally a particle.
   (B) All objects can exhibit either wave-like properties or particle-like properties depending on the situation.
   (C) Whether an interference pattern can be observed depends on the object's de Broglie wavelength.
   (D) All objects can be made to exhibit an interference pattern under the right circumstances.

21. For an RC circuit, which of the following is the correct analogy for the instantaneous current behavior in a pathway containing a capacitor? Choose the option that would give the correct instantaneous current values in the capacitor's path.

	Uncharged Capacitor	Fully Charged Capacitor
(A)	Closed switch	Open switch
(B)	Open switch	Closed switch
(C)	Closed switch	Closed switch
(D)	Open switch	Open switch

Section II: Free-Response

> **DIRECTIONS:** Answer all parts of both free-response questions using complete sentences.

1. Use the following circuit to answer the questions below.

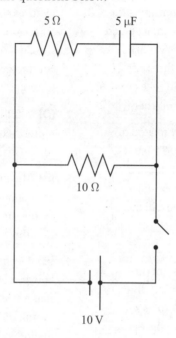

(a) When the capacitor is still uncharged and the switch is initially closed, how much current goes through the battery? Explain your reasoning.

(b) After the capacitor has been fully charged, how much current is being supplied by the battery? Explain your reasoning.

(c) After the capacitor has been fully charged, the battery is removed, leaving an open circuit where the battery was just located. Determine:

   i. The direction of current through the 10-ohm resistor

   ii. The total amount of charge that passes through the 10-ohm resistor

   iii. The initial instantaneous current after the battery has been removed

   iv. The approximate time for the capacitor to be half-discharged

2. A student is provided with a solenoid attached to an ammeter and with a bar magnet. The student monitors the ammeter as she performs the following sequence of events. Describe qualitatively her observation of the ammeter (comparing current sign and magnitude) at each step of her procedure.

(a) The student slowly inserts the north pole of the bar magnet into the solenoid.

(b) She holds the magnet steady at this location.

(c) The student withdraws the magnet quickly from the solenoid.

(d) She repeats the steps but uses the south pole of the magnet.

What factors about this procedure and the equipment affect the magnitude of current observed?

Answer Explanations

Section I

1. **(B)** The correct sequence is radiative → conductive → convective. Radiative transfer is the transfer of energy via electromagnetic radiation (such as sunlight). Conductive transfer is the transfer of thermal energy via direct contact (such as a floating object in contact with the water). Convective transfer is the transfer of thermal energy due to the movement of a fluid (like currents in the water). [Chapter 1, Thermal energy]

2. **(D)** Average kinetic energy is directly related to temperature. Since both gases are at STP (standard temperature and pressure), their average kinetic energies are the same. However, since helium has a lower mass, it must have a higher velocity to obtain the same kinetic energy ($\frac{1}{2}mv^2$). [Chapter 1, Kinetic-molecular theory]

3. **(D)** The term *isobaric* indicates constant pressure, so choice (A) is false. During path *A*, the gas expands and therefore does work to its surroundings. So choice (B) is false. Over the entire process, the enclosed area in the diagram is the work done by the gas. Since this area does not equal zero, choice (C) is false. Path *B* involves an expansion or contraction (displacement) of the gas. Therefore, no work is done during this part of the process. [Chapter 1, *P-V* diagram]

4. **(C)** Remember kq_1q_2/r^2. The charges (q_1 and q_2) are the same, and k is a constant. So the only variation in the two situations is the average r between the spheres. In the case of the two smaller spheres, the average distance (represented by the centers of the spheres) is actually closer. Therefore, the force is stronger in that case. [Chapter 2, Coulomb's law]

5. **(D)** When the spheres contact each other, electrons move off of the smaller sphere to the larger sphere and attempt to neutralize both spheres. Since this is a net +2 microcoulombs of charge, the electrons distribute themselves over both surface areas of the spheres, keeping the positive charges as far from each other as possible. Since the larger sphere has more surface area, it captures a greater portion of the excess charge than does the smaller sphere. [Chapter 2, Charge transfer]

6. **(C)** Field lines should point away from the positive and toward the negative, which all the pictures show. Field lines should never cross, and none of these diagrams has that. Field lines should also be perpendicular to the surface of conductors. Picture (C) violates this principle. [Chapter 2, Electric fields near a conductor]

7. **(D)** A fully charged capacitor cannot accept any more charge; therefore, the current going into it is zero. [Chapter 3, RC circuits]

8. **(B)** The meter pictured is a voltmeter. The voltage supplied by the battery will be split evenly among the three identical resistors. [Chapter 3, Kirchhoff's Rules]

9. **(B)** Electron configuration is the difference between a nonmagnetic and a magnetic material. Any magnetic material can be made to have overall north and south poles by lining up all of the individual magnetic domains within the material. (Each domain has its own north and south poles.) Magnetic poles do not exist in isolation as all magnetic field lines loop back on themselves. [Chapter 4, Magnetic materials]

10. **(C)** The moving charges in the wire create loops of magnetic field around them. As R increases, the field strength decreases proportional to $1/R$ rather than $1/R^2$ because the source of the field is a line rather than a point. [Chapter 4, Magnetic fields]

11. **(D)** Reversing the direction of one current will switch the direction of its magnetic field. However, when using the right-hand rule and since the direction of the moving charges within the reversed field is also reversed, the same direction of force is found. Doubling the source current will double the field strength, which then is acting on twice as much current and creating four times the force.

$$B \propto I_1$$
$$F \propto I_2 B$$
$$F \propto I_1 I_2$$

[Chapter 4, Magnetic forces]

12. **(D)** Only while the flux of the field through the loop is changing will there be an induced emf causing the current. The induced current will be such that its magnetic field will oppose the change. Since the magnetic field flux upward through the loop is getting smaller as the loop is pulled out, the induced current will create an additional upward magnetic field. The right-hand rule for current indicates a counterclockwise current is needed. [Chapter 4, Lenz's law]

13. **(D)** When a charge is accelerating, the changing velocity of the charge creates a changing magnetic field. The changing magnetic field induces a changing electric field. This dual oscillation is self-perpetuating and is known as electromagnetic radiation (or light when the frequency of oscillation lies within the visible range). [Chapter 4, Electromagnetic induction]

14. **(B)** Since the second medium is slower, it will have a larger n value, thereby requiring a smaller $\sin \theta$ value. So a smaller angle is needed:

$$n_1 \sin \theta_1 = n_2 \sin \theta_2$$

[Chapter 6, Snell's law]

15. **(D)** This is a situation of total internal reflection. Since 50 degrees is beyond the critical angle for this pair of materials, all of the light will be reflected and none will be refracted. Remember that a 90-degree "refraction" happens right at the critical angle. [Chapter 6, Total internal reflection]

16. **(A)** Convex mirrors are diverging. This means the real, reflected rays do not come to an actual focus. The focus is virtual and located behind the mirror. So any detection equipment placed at the focus would not actually receive a signal! [Chapter 6, Reflection/mirrors]

17. **(C)**

$$D \sin \theta = m\lambda$$

Wavelength does not change. So as D gets smaller, $\sin \theta$ must increase. Therefore, the locations of the bright, constructive interference fringes (integer m's) are found at greater angles. [Chapter 5, Diffraction patterns]

18. **(C)** Since the nucleus has emitted a photon of electromagnetic energy, the net charge on the nucleus must remain the same. Since energy left the nucleus, the mass of the nucleus as a whole must be decreased by E/c^2. [Chapter 7, Nuclear reactions]

19. **(D)** Light is emitted (bright spectral line) when an electron falls from an excited state to a lower energy state. The greater the energy *difference* is (as opposed to the starting energy state), the higher the frequency of light emitted ($E = hf$). [Chapter 7, Light/matter interactions]

20. **(C)** The de Broglie wavelength is the connection between classically particle-like objects and their wave-like properties. Interference patterns have been successfully shown for many particles. Macroscopic objects, however, have such large masses that their de Broglie wavelengths make observing a diffraction pattern impossible:

$$\lambda = h/p$$

[Chapter 7, Wave-particle duality]

21. **(A)** An uncharged capacitor will initially appear to be a simple wire as the first charge to hit one side of the capacitor draws its opposite on the other side. On the other hand, a fully charged capacitor accepts no additional charges and blocks any current from entering its pathway. These two states are connected by an exponential function:

$$I = (\Delta V/R)\, e^{-t/RC}$$

Remember to compare $t = 0$ to $t \gg RC$. [Chapter 3, Capacitors]

Section II

1. This is an electric circuit problem. [See Chapter 3 for more information]

   (a) Initially the top path acts as a simple 5-ohm resistor because the capacitor is uncharged and will act as if current is passing right through it. The 10-ohm resistor is in parallel with the capacitor, resulting in an equivalent resistance of 10/3 ohms:

   $$1/5 + 1/10 = 1/R$$
   $$2/10 + 1/10 = 1/R$$
   $$3/10 = 1/R$$
   $$R = 3\,\Omega$$

   Ohm's law gives us a 3-amp current coming from the battery:

   $$\Delta V = IR$$
   $$10 = I\,(10/3)$$
   $$I = 3\,\text{A}$$

   (b) After the capacitor has been fully charged, it will act as an open circuit, no longer allowing charge to flow through the upper path. Equivalent resistance for the circuit will now be 10 ohms:

   $$\Delta V = IR$$
   $$10\,\text{V} = I\,(10\,\Omega)$$
   $$I = 1\,\text{A}$$

   (c) i. The direction of current is constant throughout the problem, from right to left, whether the current is being supplied by the battery or by the capacitor.

       ii.

   $$C = Q/V$$
   $$50\,\mu\text{F} = Q/10\,\text{V}$$

   The entire 10 V is on the capacitor at the end because the 5-ohm resistor is not using any voltage ($I = 0$):

   $$Q = 50\,\mu\text{C}$$

       iii. Initially, the 10 volts of the capacitor will be driving current through an equivalent resistance of 15 ohms ($5 + 10$ in series):

   $$V = IR$$
   $$10\,\text{volts} = I\,(15\,\text{ohms})$$
   $$I = 0.67\,\text{amps}$$

   Note that right away, the voltage will begin to decrease across the capacitor as it discharges, therefore driving increasingly smaller currents through the resistors.

       iv. As an exponential decreasing function, the time constant of RC provides an approximate time for losing about half the charge:

   $$RC = (15\,\text{ohms})(5\,\text{microfarads})$$
   $$= 75\,\text{microseconds}$$

2. This is a magnetism and electromagnetism problem. [See Chapter 4 for more information]

   (a) The student will see a modest current in one direction as the bar magnet enters the solenoid. Note that without knowing the actual orientation of the loops, the magnet, and the ammeter, the actual sign of the current cannot be determined.

   (b) She will observe no current while the magnet is being held still.

   (c) The student will observe a current in the opposite direction as it is withdrawn. The magnitude of this current will be larger as the change in magnetic flux is happening quicker.

   (d) The same as described above but with opposite signs for the current.

   The number of coils, the total resistance of the coils, the strength of the magnet, the speed at which the magnet is moved, and possibly the orientation of the magnet relative to the cross-sectional area of the solenoid all affect the magnitude of the observed current.

Review and Practice

1

Thermodynamics

Learning Objectives

In this chapter, you will learn:

- → Temperature and its measurement
- → Molar quantities
- → The ideal gas law
- → Kinetic-molecular theory
- → Work done by expanding gases
- → The first law of thermodynamics
- → The second law of thermodynamics and heat engines
- → Heat transfer
- → Specific heat

Temperature and Its Measurement

Temperature can be defined as the relative measure of heat. In other words, temperature defines what will lose and what will gain thermal energy when the energy is transferred. We can state even more simply that temperature is a measure of how "hot" or "cold" a body is (relative to your own body or another standard or arbitrary reference). The source of the temperature of an object is, in actuality, the average kinetic energy of its atoms. Human touch is unreliable in measuring temperature since what feels hot to one person may feel cool to another.

During the seventeenth century, Galileo invented a method for measuring temperature using water and its ability to react to changes in heat content by expanding or contracting. Figure 1.1 shows a simple "thermometer" that consists of a beaker of water in which there is an inverted tube with a circular bulb at the top. When thermal energy is added to the beaker, the level of water will rise.

TIP

Heat and temperature are not the same thing.

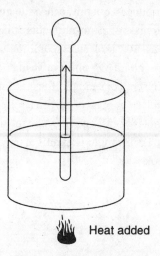

Heat added

Figure 1.1 Thermometer

For two reasons, however, water is not the best substance to use to measure temperature. First, water has a relatively high freezing point and a low boiling point; second, it has a peculiar variation in density as a function of temperature because of its chemical structure. Even so, the phenomenon of "thermal expansion" is an important topic not only in this context but also for engineering buildings, roads, and bridges. We will discuss this aspect in more detail later in the chapter.

Most modern thermometers (see Figure 1.2) use mercury or alcohol in a small capillary tube. In the metric system, the standard sea-level points of reference are the freezing point (0 degrees) and boiling point (100 degrees) of water. This temperature scale, formerly known as the centigrade scale, is now called the Celsius scale after Anders Celsius. (The English Fahrenheit system is not used in physics.)

Figure 1.2 Thermometer

One difficulty with the Celsius scale is that the zero mark is not a true zero since a substance can be much colder than zero. Some gases, such as oxygen and hydrogen, do not even condense until their temperatures reach far below 0 degrees Celsius. The discovery of a theoretical "absolute zero" temperature resulted in the so-called absolute or Kelvin temperature scale. A temperature of −273 degrees Celsius is redefined as "absolute zero" on the Kelvin scale. No temperature colder than absolute zero is possible since all atomic motion ceases and there is no more kinetic energy to produce a temperature. An easy way to convert from Celsius to Kelvin is to add the number 273 to the Celsius temperature. This just adds a scale factor that preserves the metric scale of the Celsius reading. In other words, there is still a 100-degree difference in temperature between melting ice and boiling water: 273 K and 373 K, respectively. (Note that neither the word "degrees" nor the degree sign is used in expressing Kelvin temperatures.)

> Note that it is not actually possible to cool a real object all the way down to absolute zero. This is known as the third law of thermodynamics or Nernst's theorem.

Molar Quantities

Under the conditions of STP (standard temperature [0°C] and pressure [1 atmosphere]), most gases behave according to a simple relationship among their pressures, volumes, and absolute temperatures. This **equation of state** applies to what is referred to as an **ideal gas**. A real gas, however, is very complex, and a large number of variables are required to understand all of the intricate movements taking place within it.

To help us understand ideal gases, we introduce a quantity called the **molar mass**. In thermodynamics (and chemistry), the molar mass or **gram-molecular mass** is defined to be the mass of a substance that contains 6.02×10^{23} molecules (commonly referred to as a mole). This number is called **Avogadro's number** (N_O), and 1 mole of an ideal gas occupies a volume of 22.4 liters at zero degrees Celsius. For example, the gram-molecular mass of carbon (C) is 12 grams, while the gram-molecular mass of oxygen (O_2) is 32 grams. If m represents the actual mass, the number of moles of gas is given by the relationship $n = m/M$, where M equals the gram-molecular weight. For example, 64 grams of oxygen equals 2 moles, and 36 grams of carbon equals 3 moles.

REMEMBER

Specifically, an ideal gas is one

1. That has only elastic collisions
2. That has no intermolecular forces and where gravity can be ignored
3. In which the molecules occupy a volume much smaller than the total volume available

The Ideal Gas Law

Imagine that we have an ideal gas in a sealed, insulated chamber such that no heat can escape or enter. An ideal gas is one in which the volume of the molecules are an ignorable fraction of the total volume of the gas and one in which the molecules do not interact with each other except via their elastic collisions. Figure 1.3 shows an external force placed on the piston at the top of the chamber so that the pressure change causes the piston to move down. The chamber is a cylinder with cross-sectional area A and height h. If a constant force F is applied, work is done to move the piston (frictionless) a distance h. Thus, $W = Fh$, and since $h = V/A$, the work done is

$$W = \frac{FV}{A} = PV$$

In other words, the work done (in this case) is equal to the product of the pressure and the volume.

If the temperature of the gas remains constant, this relationship is known as **Boyle's law** and can be expressed as

$$PV = \text{constant}$$

or

$$P_1 V_1 = P_2 V_2$$

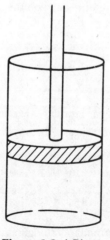

Figure 1.3 A Piston

Experiments demonstrate that, if a gas is enclosed in a chamber of constant volume, there is a direct relationship between the pressure in the chamber and the absolute Kelvin temperature of the gas: the ratio of pressure to absolute temperature in an ideal gas always remains constant (indicated by a diagonal, straight-line graph at constant volume). We can write this relationship algebraically as:

$$\frac{P}{T} = \text{constant}$$

or

$$\frac{P_1}{T_1} = \frac{P_2}{T_2}$$

where the temperatures must be in kelvins.

TIP

An *ideal gas* is one made of pointlike particles that collide elastically.

French scientist Jacques Charles (and, independently, Joseph Gay-Lussac) developed a relationship, commonly known as **Charles's law**, between the volume and the absolute temperature of an ideal gas at constant pressure. This law states that the ratio of volume and absolute temperature in an ideal gas remains constant; that is,

$$\frac{V_1}{T_1} = \frac{V_2}{T_2}$$

The three gas laws—pressure-temperature law, Boyle's law, and Charles's law—can be summarized by the one **ideal gas law**, which has four forms:

The first form is a direct consequence of all three gas laws:

$$\frac{P_1 V_1}{T_1} = \frac{P_2 V_2}{T_2}$$

where the temperatures must be in kelvins.

The second form states that the product of pressure and volume, divided by the absolute temperature, remains a constant:

$$\frac{PV}{T} = \text{constant}$$

The third and more general form of the ideal gas law (and the one found in most advanced textbooks) includes the number of moles of gas involved and a universal gas constant R:

$$PV = nRT$$

where $R = 8.31$ joules per mole $\cdot$ kelvin (J/mol $\cdot$ K).

Finally, if $n = N/N_O$, where N is the actual number of molecules and N_O is Avogadro's number, a constant $k_B = R/N_O$, called **Boltzmann's constant**, can be introduced, and the ideal gas law takes the form

TIP

The state of an ideal gas is uniquely specified by its pressure, volume, and temperature.

$$PV = NkT$$

Again, the temperatures must always be in kelvins.

> **Sample Problem 1**

A confined gas at constant temperature has a volume of 50 m³ and a pressure of 500 Pa. If it is compressed to a volume of 20 m³, what is the new pressure?

> **Solution**

We use Boyle's law:

$$P_1 V_1 = P_2 V_2$$
$$(500 \text{ Pa})(50 \text{ m}^3) = P_2 (20 \text{ m}^3)$$
$$P_2 = 1,250 \text{ Pa}$$

> **Sample Problem 2**

A confined gas is at a temperature of 27°C when it has a pressure and volume of 1,000 Pa and 30 m³, respectively. If the volume is decreased to 20 m³ and the temperature is raised to 30°C, what is the new pressure?

> **Solution**

We will use the ideal gas law, but we must remember that the temperatures must be in Kelvin!

$$27°\text{C} = 300 \text{ K}$$
$$30°\text{C} = 303 \text{ K}$$

Now,

$$\frac{(P_1 V_1)}{T_1} = \frac{(P_2 V_2)}{T_2}$$
$$\frac{(1,000 \text{ Pa})(30 \text{ m}^3)}{300 \text{ K}} = \frac{P_2 (20 \text{ m}^3)}{303 \text{ K}}$$

Solving, we obtain

$$P_2 = 1,515 \text{ Pa}$$

Kinetic-Molecular Theory

What is thermal or internal energy? When thermal energy is exchanged, we call this heat. If an object is losing thermal energy, its total internal (or thermal) energy decreases. This can manifest in either a lowering of temperature or a phase change. Temperature is a proxy for microscopic kinetic energy. The energies associated with phase changes are microscopic potential energies. By microscopic, we mean the kinetic energies associated with the individual molecular velocities and the potential energies of relationships between the molecules within the substance. Kinetic-molecular theory refers to the fact that if an object remains in its current state, any thermal energy is mostly going to or coming from the kinetic energy of the individual molecules or atoms within the substance. Specifically for an ideal gas, all thermal energy is in the form of molecular kinetic energy. Recall that one of the assumptions for an ideal gas is that the molecules are not interacting and have perfectly elastic collisions:

Average kinetic energy per molecule $= (3/2)kT$

where k is the Boltzmann constant and T is the temperature in Kelvin. Therefore, the total thermal energy of an ideal gas is $(3/2)NkT$, where N is the number of molecules in the substance. Some molecules have higher speeds and thus higher kinetic energies. Some will have lower kinetic energies. However, the average is given by the formula above.

If one uses the kinetic energy formula $((1/2)mv^2)$ and solves for velocity, an expression is obtained for the root-mean-square (RMS) velocity:

$$v_{rms} = \sqrt{\frac{3kT}{m}}$$

Note that this does not represent the average speed of the molecules. RMS values are bigger than the actual mean speed. The exact relationship depends on the distribution of speeds. It is the Maxwell-Boltzmann distribution that is most commonly used for idealized gases. You will not be expected to have detailed knowledge of this distribution for the AP Physics 2 exam, but the facts presented above and the fact that the distribution has a broad distribution of speeds is important to know.

> ### Sample Problem 3

An ideal gas at 120 degrees Celsius has molecules with an average kinetic energy of what value? How many molecules would you need to have 0.5 joules of thermal energy?

✓ ### Solution

$$T = 120 + 273 = 393 \text{ Kelvin}$$
$$KE_{average} = (3/2)kT = (3/2)(1.38 \times 10^{-23})(393) = 8.14 \times 10^{-21} \text{ joules}$$
$$\text{Number of molecules} = 0.5 \text{ J}/(8.14 \times 10^{-21} \text{ J}) = 6.14 \times 10^{19} \text{ molecules}$$

Work Done by Expanding Gases

If a gas expands by displacing any substance, the gas is doing positive work to its environment. Therefore, the gas is transferring energy out of its internal energy into the environment. By the kinetic-molecular theory, this internal energy is in the kinetic energy of the molecules. This loss of energy, therefore, means its temperature will go down as the molecules slow down. Think of each individual collision of a molecule at the edge of the gas with its surroundings. In order to displace the external molecule it hits (the expansion part), the gas molecule must do some individual work and, in doing so, lower its own kinetic energy. As each gas molecule that hits the exterior goes through this, the average speed is lowered, which lowers the temperature of the entire gas.

Note that pressure times volume has units of energy. For expanding gases, the area under a pressure-versus-volume curve is the work done by the gas (representing a loss of energy from the gas). For contracting gases, the area is the work being done to the gas (representing a gain in energy to the gas). If a gas is cycled through compression and expansion, the difference in areas represents the net work done by the gas.

> **Sample Problem 4**

If 2 m³ of gas at 2,000 Pa undergoes isobaric (constant pressure) compression to 1 m³, then is heated while held at that size such that its pressure is increased to 5,000 Pa, and then is allowed to undergo isobaric expansion to 3 m³, what is the net work done to the gas?

✓ **Solution**

During the compression, the area under the curve is the rectangle represented by $P\Delta V = 2{,}000(2 - 1\ m^3) = 2{,}000$ J of work done **to** the gas. The heating at constant volume represents no work done as the area under the vertical line is zero. The final expansion represents work done **by** the gas of $P\Delta V = 5{,}000(3 - 1\ m^3) = 10{,}000$ J.

Net work done **to** the gas $= +2{,}000$ J $- 10{,}000$ J $= -8{,}000$ J.

In other words, the gas did $+8{,}000$ J of work to its surroundings.

> **Sample Problem 5**

Twenty cubic meters of an ideal gas is at 50°C and at a pressure of 50 kPa. If the temperature is raised to 200°C and the volume is compressed by half, what is the new pressure of the gas?

✓ **Solution**

We must first change all temperatures to kelvins:

$$50°C = 323\ K \quad \text{and} \quad 200°C = 473\ K$$

We now use the ideal gas law for two sets of conditions:

$$\frac{P_1 V_1}{T_1} = \frac{P_2 V_2}{T_2}$$

Substituting the appropriate values, we get

$$\frac{(50)(20)}{323} = \frac{(P_2)(10)}{473}$$

Solving for the pressure gives $P_2 = 146.4$ kPa.

The First Law of Thermodynamics

If work is being done to an ideal gas at constant temperature, Boyle's law states that the work done is equal to the product of the pressure and the volume (this product remaining constant). A graph of Boyle's law, seen in Figure 1.4, is a hyperbola. Pressure and volume changes that include temperature changes and are adiabatic obey a different kind of inverse relationship, in which the volume is raised to a higher exponential power (making the hyperbola steeper).

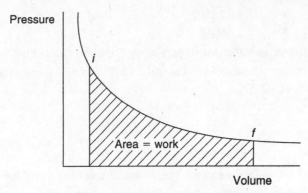

Figure 1.4 Work in a *P-V* Diagram

The pressure-versus-volume diagram for Boyle's law demonstrates the general fact that the area enclosed by the curve from some initial state to a final state is equal to the work done.

The **first law of thermodynamics** is basically the law of conservation of energy. Let Q represent the change in heat energy supplied to a system, ΔU represent the change in internal molecular energy, and W represent the amount of net work done on the system. The work done by a gas is its pressure times change in volume. Then the first law of thermodynamics states that

$$\Delta U = Q + W$$

REMEMBER

Sign Conventions:

Q is "+" for thermal energy coming into the gas and "−" for thermal energy leaving the gas.

W is "+" for work done to the gas and "−" for work done by the gas to its surroundings.

In thermodynamics, certain terms are often used and worth examining in the case of an ideal gas:

Isothermal	Isobaric	Isochoric	Adiabatic
Constant temperature	Constant pressure	Constant volume	No heat transfer
P–*V* curve	*P*–*V* curve	*P*–*V* curve	*P*–*V* curve
$\Delta U = 0$	$W = P\Delta V$	$W = 0$	$Q = 0$

The Second Law of Thermodynamics and Heat Engines

No system is ideal. When a machine is manufactured and operated, heat is always lost because of friction. This friction creates heat that can be accounted for by the law of conservation of energy. Recall that heat flows from a hotter body to a cooler body. Thus, if there is a reservoir of heat at high temperature, the system will draw energy from that reservoir (hence its name). Only a certain amount of energy is needed to do work. The remainder is released as exhaust at a lower temperature. This phenomenon was investigated by a French physicist named Sadi Carnot. Carnot was interested in the cyclic changes that take place in a heat engine, which converts thermal energy into mechanical energy.

In the nineteenth century, German physicist Rudolf Clausius coined the term *entropy* to describe an increase in the randomness or disorder of a system. These studies by Clausius and Carnot can be summarized as follows:

1. Heat never flows from a cooler body to a hotter body of its own accord.
2. Heat can never be taken from a reservoir without something else happening in the system.
3. The entropy of any isolated system is always increasing.

These three statements embody the **second law of thermodynamics**: it is impossible to construct a cyclical machine that produces no other effect than to transfer heat continuously from one body to another at higher temperature.

Connected with the second law of thermodynamics is the concept of a reversible process. Heat flows spontaneously from hot to cold, and this action is irreversible unless acted upon by an outside agent. A system may be reversible if it passes from the initial state to the final state by way of some intermediary equilibrium states. This reversible process, which can be represented on a pressure-versus-volume diagram, was studied extensively by Carnot.

A simple heat engine is shown in Figure 1.5.

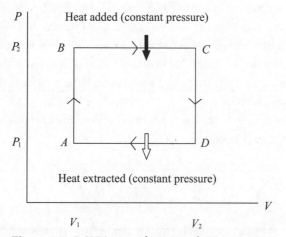

Figure 1.5 *P-V* Diagram for a Simple Heat Engine

From point A to point B, pressure increases at constant volume. No work is done. From point B to point C, heat is absorbed from the outside as the gas expands at constant pressure. The work done is equal to $P_2(V_2 - V_1)$. From point C to point D, the volume is constant as the pressure decreases. No work is done. Finally, from point D to point A, the volume decreases at constant pressure and heat is lost to the outside. The work done is equal to $P_1(V_1 - V_2)$.

In each isochoric change, $\Delta W = 0$ since there were no changes in volume. During isobaric changes,

$$\Delta W_2 = P_2(V_2 - V_1) \quad \text{and} \quad \Delta W_1 = P_1(V_1 - V_2)$$

The net work is equal to $(P_2 - P_1)(V_2 - V_1)$ and would be equal to the area of the rectangle in Figure 1.6.

The Carnot cycle is one in which adiabatic temperature changes occur, as shown in Figure 1.6.

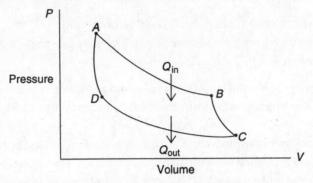

Figure 1.6 Carnot Cycle

The isothermal changes occur from A to B and from C to D. The adiabatic changes are represented by segments BC and AD, in which no heat is gained or lost. Therefore, heat must be added in segment AB and released in CD. In other words, there is an isothermal expansion in AB at a temperature T_1. Heat is absorbed from the reservoir, and work is done. BC represents an adiabatic expansion in which an insulator has been placed in the container of the gas. The temperature falls, and work is done. In CD, we have an isothermal compression when the container is connected to a reservoir of temperature T_2 that is lower than T_1. Heat is released in this process, and work is done. Finally, in DA, the gas is compressed adiabatically, and work is done. The total work done would be equal to the area of the figure enclosed by the graph.

The efficiency of heat engines is given by the fractional ratio between the total work done and the original amount of heat added (since $\Delta U = 0$ for the cycle, $W = Q_{in} - Q_{out} = Q_1 - Q_2$):

$$\text{Efficiency} = \frac{W}{Q_1} = 1 - \frac{Q_2}{Q_1}$$

The reason is that the net work done in one cycle is equal to the net heat transferred $(Q_1 - Q_2)$ since the change in internal energy is zero. For the Carnot cycle, this can be rewritten as

$$\text{Efficiency} = 1 - \frac{T_2}{T_1}$$

where $T_1 > T_2$ and both temperatures are in kelvins.

A heat pump, such as those used by refrigerators, can be thought of as reversed Carnot cycles. The same steps are executed but in the opposite sense. Work is done *to* the fluid rather than *by* the fluid, and the difference in heat added and extracted over the cycle is negative such that more heat is extracted from the fluid than is added.

> **Sample Problem 6**

A confined gas undergoes changes during a thermodynamic process as shown in the *P-V* diagram. Calculate the net energy added as heat during one cycle.

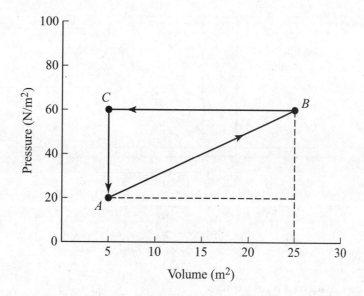

✓ **Solution**

We can calculate the net energy by calculating the area *under* each segment of the graph. From $A \rightarrow B$, the work done *below* the diagonal segment is equal to 800 J. From $B \rightarrow C$, the area below the straight-line segment is equal to 1,200 J. No work is done from $C \rightarrow A$ since we do not have a change in volume. Thus, the net energy is $800 \text{ J} - 1{,}200 \text{ J} = -400 \text{ J}$.

Heat Transfer

Heat can be transferred by conduction, convection, or radiation.

Because metals contain a large number of free electrons, heat **conduction** is the main mode of heat transfer. If you have ever left a spoon in a cup of hot water for a long time, you have experienced heat conduction.

The rate of heat transfer ($\Delta Q/t$) through a solid slab of material (see Figure 1.7) depends on four quantities:

1. The temperature difference on either side of the slab (ΔT).
2. The thickness of the slab (L).
3. The frontal area of the slab (A).
4. The thermal conductivity (k) of the material. This is a number in units of watts per meter per degree Celsius (W/m · °C).

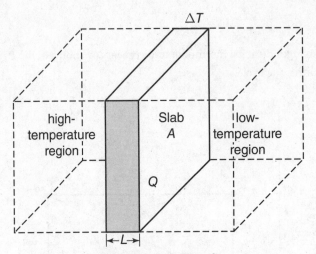

Figure 1.7 Heat Transfer

$$H = \frac{\Delta Q}{t} = \frac{kA\Delta T}{L}$$

Table 1.1 lists some typical thermal conductivities.

Table 1.1 Thermal Conductivities

Material	$k(W/m \cdot °C)$
Air	0.024
Aluminum	2.50
Brick	0.6
Brass	109
Concrete	0.8
Cork	0.04
Copper	385
Glass	0.8
Ice	1.6
Iron, steel	50
Silver	406

Heat can be transferred by **convection** in a gas or liquid as matter circulates throughout the system. For example, hot air is less dense and rises while cold air is more dense and sinks.

If the air above the ground is heated by the Sun, that air will rise and cool. This colder air will sink, and circulation will occur. A room can be heated by convection as hot air displaces cold air and circulates throughout the room. Convection currents can also be seen in water being heated. If a dye is placed into water that is being heated from below, the patterns formed will show the flow of convection.

If heat is transferred by means of electromagnetic waves (see Chapter 5), the process is known as **radiation**. Infrared waves from the Sun are a typical mechanism for heating the atmosphere. Radiation, like all waves, transfers energy only from place to place. Radiation does not require a material medium for propagation.

Specific Heat

Unless it is during a phase transition, as heat (Q) is transferred into or out of an object, its temperature will go up or down, respectively. The relationship between heat absorbed and the temperature increase is dependent on two aspects of the object: its mass (m) and its specific heat (c). The nature of the atoms and the way they interact determines the specific heat of that substance (therefore, even the same compound will have different values for specific heat in its different phases). The most common substance that is used to measure heat is liquid water. Water has a specific heat of 4.18 J/(g · K). This is actually an uncommonly high value for specific heat and thus makes water very efficient at holding thermal energy (requiring lots of energy in order to raise its temperature moderately). Recall that temperature *changes* can be in either degrees Celsius or Kelvin but never Fahrenheit. Below is the formula for determining the heat required, for an object of a given mass and specific heat, to undergo a certain change in temperature:

$$Q = mc\Delta T$$

SUMMARY

- Pressure is defined as the force per unit area.
- One mole of an ideal gas occupies a volume of 22.4 liters and contains 6.02×10^{23} molecules (Avogadro's number) at STP.
- Boyle's law states that if a confined ideal gas is kept at constant temperature, then the product of the pressure and the volume remains constant (PV = constant).
- Charles's law states that if a confined ideal gas is kept at constant pressure, then there is a direct relationship between the volume and absolute temperature (V/T = constant).
- For an ideal gas, the relationship PV/T is equal to a constant.
- The first law of thermodynamics is essentially a restatement of the law of conservation of energy: $\Delta U = \Delta Q + \Delta W$.

Thermodynamic Sign Conventions		
Q	**W**	**ΔU**
+	Heat into fluid	Work done to fluid	Fluid temperature increases
−	Heat out of fluid	Work done by fluid	Fluid temperature decreases

- The second law of thermodynamics relates to an increase in entropy (randomness) in the universe and the fact that heat is always generated when work is done. It is therefore impossible to create an ideal machine without any losses due to heat.
- Ideal gases can be described using a statistical theory known as the kinetic theory of gases in which the molecules of a gas undergo elastic collisions and the temperature is a measure of the average kinetic energy of the molecules.
- Heat can be transferred by conduction, convection, or radiation.

Problem-Solving Strategies for Thermodynamics and the Kinetic Theory of Gases

It is important to remember that all temperatures must be converted to kelvins before doing any calculations. Also, since the units for the gas constant R and Boltzmann's constant k are in SI units, all masses or molecular masses must be in kilograms. For example, the molecular mass of nitrogen is usually given as 28 grams per mole. However, in an actual calculation, this must be converted to 0.028 kilograms per mole.

You also should remember that, unless a comparison relationship is used (such as Boyle's law), all pressures must be in units of pascals (not kilopascals, atmospheres, or centimeters of mercury) and all volumes must be in cubic meters. However, in a comparison relationship, the chosen units are irrelevant (except for temperature, which is always in kelvins), and you can use atmospheres for pressure, or liters (1 liter = 1,000 cubic centimeters) for volume. The molar specific heat capacities depend on the number of moles, not the number of grams (the specific heat capacity).

Practice Exercises

1. At constant temperature, an ideal gas is at a pressure of 30 cm of mercury and a volume of 5 L. If the pressure is increased to 65 cm of mercury, the new volume will be

    (A) 10.8 L.
    (B) 2.3 L.
    (C) 1.7 L.
    (D) 0.43 L.

2. At constant volume, an ideal gas is heated from 75°C to 150°C. The original pressure was 1.5 atm. After heating, the pressure will be

    (A) doubled.
    (B) halved.
    (C) the same.
    (D) less than doubled.

3. At constant pressure, 6 m³ of an ideal gas at 75°C is cooled until its volume is halved. The new temperature of the gas will be

    (A) 447°C.
    (B) 174°C.
    (C) 37.5°C.
    (D) −99°C.

4. Water is used in an open-tube barometer. If the density of water is 1,000 kg/m³, what will be the level of the column of water at sea level?

    (A) 10.3 m
    (B) 11.2 m
    (C) 12.5 m
    (D) 13.6 m

5. As the temperature of an ideal gas increases, the average kinetic energy of its molecules

    (A) increases, then decreases.
    (B) decreases.
    (C) remains the same.
    (D) increases.

6. The product of pressure and volume is expressed in units of

    (A) pascals.
    (B) kilograms per newton.
    (C) watts.
    (D) joules.

7. Which of the following is equivalent to 1 Pa of gas pressure?

    (A) 1 kg/s^2
    (B) $1 \text{ kg} \cdot \text{m/s}$
    (C) $1 \text{ kg} \cdot \text{m}^2/\text{s}^2$
    (D) $1 \text{ kg/m} \cdot \text{s}^2$

8. What is the efficiency of a heat engine that performs 700 J of useful work from a reservoir of 2,700 J?

    (A) 26%
    (B) 35%
    (C) 65%
    (D) 74%

9. The first law of thermodynamics is a restatement of which law?

    (A) conservation of charge
    (B) conservation of energy
    (C) conservation of entropy
    (D) conservation of momentum

10. Which of the following graphs represents the relationship between pressure and volume for an ideal confined gas at constant temperature?

(A)

(B)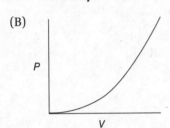

(C)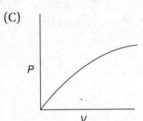

(D)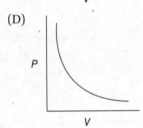

11. (a) How many molecules of helium are required to fill a balloon with a diameter of 50 cm at a temperature of 27°C?

(b) What is the average kinetic energy of each molecule of helium?

(c) What is the average velocity of each molecule of helium?

12. A heat engine absorbs 2,000 J of heat from a hot reservoir and expels 750 J to a cold reservoir during each operating cycle.

(a) What is the efficiency of the engine?

(b) How much work is done during each cycle?

(c) What is the power output of the engine if each cycle lasts for 0.5 s?

13. A monatomic ideal gas undergoes a reversible process as shown in the pressure-versus-volume diagram below:

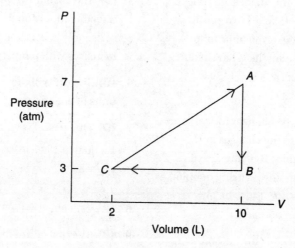

(a) If the temperature of the gas is 500 K at point A, how many moles of gas are present?

(b) For each cycle (AB, BC, and CA), determine the values of ΔW, ΔU, and ΔQ.

14. A glass window has dimensions of 1.2 m high, 1.0 m wide, and 5 mm thick. The outside temperature is 5°C, while the inside air temperature is 15°C.

(a) What is the conductive rate of heat transfer through the window?

(b) How much energy is transferred in one hour?

Answer Explanations

1. **(B)** At constant temperature, an ideal gas obeys Boyle's law. Since we are making a two-position comparison, we can use the given units for pressure and volume. Substituting the given values into Boyle's law, we obtain $(30)(5) = (65)(V_2)$, which implies that $V_2 = 2.3$ L.

2. **(D)** At constant volume, the changes in pressure are directly proportional to the changes in absolute Kelvin temperature. Therefore, even though the Celsius temperature is being doubled, the Kelvin temperature is not! Thus, the new pressure must be less than doubled.

   Alternatively, you can calculate the new pressure since $75°C = 348$ K and $150°C = 423$ K. Thus, $1.5/348 = P_2/423$, and $P_2 = 1.8$ atm (less than double). Again, since a comparison relationship is involved, we can keep the pressure in atmospheres for convenience.

3. **(D)** This is a Charles's law problem, and the temperatures must be in kelvins. Since $75°C = 348$ K, we can write that $6/348 = 3/T_2$. However, the volume change is directly proportional to the Kelvin temperature change. Thus, $T_2 = 174$ K $= -99°C$.

4. **(A)** The formula for pressure is $P = pgh$. The density of water is $1,000$ kg/m^3, and $g = 9.8$ m/s^2. We also know that at sea level $P = 101$ kPa, which must be converted to pascals or newtons per square meter. Thus, $h = P/pg$, and making the necessary substitutions gives $h = 10.3$ m (a very large barometer over 33 ft tall!).

5. **(D)** The absolute temperature of an ideal gas is directly proportional to the average kinetic energy of its molecules. Hence, the average kinetic energy increases with temperature.

6. **(D)** The units of the product PV are joules, the units of energy.

7. **(D)** One pascal is equivalent to 1 N/m^2, but 1 N is equal to 1 kg $\cdot$ m/s^2. Thus, the final equivalency is $\dfrac{1 \text{ kg}}{\text{m} \cdot \text{s}^2}$.

8. **(A)** Efficiency $= (700 \text{ J}/2{,}700 \text{ J}) \times 100\% = 26\%$.

9. **(B)** The first law of thermodynamics expresses the conservation of energy.

10. **(D)** Boyle's law relates pressure and volume in an ideal gas. This is an inverse relationship that looks like a hyperbola.

11. (a) From the ideal gas formula, we know that $PV = NkT$, where N equals the number of molecules and k is Boltzmann's constant. For a 50 cm diameter balloon, the radius will be 0.25 m. One atmosphere is equal to 101,000 Pa of pressure. The volume of a sphere is given by $V = (4/3)\pi r^3$, which equals 0.065 m³ upon substitution.

We also know that $N = PV/kT$, where T is equal to 300 K in this problem. Substituting all known values gives

$$N = \frac{(101,000)(0.065)}{(1.38 \times 10^{-23})(300)} = 1.59 \times 10^{24} \text{ molecules}$$

(b) The average kinetic energy of each molecule is equal to

$$\overline{KE} = \frac{3}{2}kT = (1.5)(1.38 \times 10^{-23})(300) = 6.21 \times 10^{-21} \text{ J}$$

(c) The average velocity of each molecule is given by

$$v_{rms} = \sqrt{\frac{3RT}{m}} = \sqrt{\frac{(3)(8.31)(300)}{0.004}} = 1,367 \text{ m/s}$$

12. (a) The efficiency of the heat engine is given by the formula

$$e = \frac{\Delta Q}{Q_{hot}} = \frac{2,000 - 750}{2,000} = 0.625 \text{ or } 62.5\%$$

(b) The work done during each cycle is equal to

$$\Delta Q = 2,000 - 750 = 1,250 \text{ J}$$

(c) The power output for each cycle equals the work done divided by the time:

$$P = \frac{1,250}{0.5} = 2,500 \text{ W}$$

13. (a) Since we know the pressure, volume, and absolute temperature of the ideal (monatomic) gas at point A, we can use the ideal gas law, $PV = nRT$, to find the number of moles, n.

First, however, we must be in standard units: 7 atm = 707,000 Pa, and since 1 L = 0.001 m³, then 10 L = 0.01 m³. Thus,

$$n = \frac{(707,000)(0.01)}{(8.31)(500)} = 1.7 \text{ mol of gas}$$

(b) i. The cycle AB is isochoric, and so

$$\Delta W = 0$$

Thus, $\Delta U = \Delta Q$. Now, as the pressure decreases from 7 atm to 3 atm, the absolute temperature decreases proportionally. Thus, we can write

$$\frac{7}{500} = \frac{3}{T_B}$$

which implies that the absolute temperature at point B is equal to 214.29 K. To find the change in the internal energy, we recall that

$$\Delta U = \frac{3}{2}nR\Delta T = (1.5)(1.7)(8.31)(-285.7)$$
$$= -6,054 \text{ J} = \Delta Q \text{ also!}$$

ii. The cycle BC is isobaric, and the decrease in volume from 10 L to 2 L is accompanied by a corresponding decrease in absolute temperature. We know from step i that the temperature at B is 214.29 K, so we can use Charles's law to find the temperature at point C. Also, the work done is negative since we are expelling energy to reduce temperature and volume. We can therefore write

$$\frac{10}{214.29} = \frac{2}{T_C}$$

which implies that the absolute temperature at point C is equal to 42.86 K. Therefore, $\Delta T = -171.43$ K.

Now, in this process, work is done to compress the gas, and $W = P(V_C - V_B)$. The pressure is constant at $P = 3$ atm $= 303,000$ Pa, and $\Delta V = -8$ L $= -0.008$ m^3.

Thus,

$$\Delta W = (303,000)(-0.008) = -2,424 \text{ J}$$

To find ΔU, we note that the temperature change of the 1.7 moles was -171.43 K, and so

$$\Delta U = (1.5)(1.7)(8.31)(-171.43) = -3,633 \text{ J}$$

To find ΔQ, we use the first law of thermodynamics:

$$\Delta Q = \Delta U + \Delta W = -3,633 - 2,424 = -6,057 \text{ J}$$

iii. The last part of the reversible cycle puts back all the work and energy, a total of 12,111 J of heat (the sum of ΔQ in steps i and ii), that was removed in the first two steps. Thus, for CA,

$$\Delta Q = +12,111 \text{ J}$$

In a similar way, the total change in internal energy equaled $-9,686.69$ J. Thus, in cycle CA,

$$\Delta U = +9,686.69 \text{ J}$$

Finally,

$$\Delta W = \Delta Q - \Delta U = 12,111 - 9,686.69 = 2,424.31 \text{ J}$$

Notice that, since no work was done in step i, this is approximately equal to the work done in step ii!

14. (a) The formula for conductive heat transfer is $H = \frac{kA\Delta T}{L}$. First, we need to change the glass thickness to meters:

$$5 \text{ mm} = 0.005 \text{ m}$$

From Table 1.1, we see that the thermal conductivity of glass is

$$k = 0.8 \text{ W/m} \cdot {}^\circ\text{C}$$

The area of the glass window is simply

$$A = (1.2 \text{ m})(1.0 \text{ m}) = 1.2 \text{ m}^2$$

By substituting in all the values and noting that $\Delta T = 10{}^\circ\text{C}$, we get

$$H = \frac{(0.8 \text{ W/m} \cdot {}^\circ\text{C})(1.2 \text{ m}^2)(10{}^\circ\text{C})}{(0.005 \text{ m})}$$
$$= 1,920 \text{ W}$$

(b) Since energy = power × time, the energy transferred in 1 hour (3,600 s) is

$$E = (1,920 \text{ J/s})(3,600 \text{ s}) = 6.912 \times 10^6 \text{ J}$$

2

Electric Force, Field, and Potential

Learning Objectives

In this chapter, you will learn:

→ The nature of electric charges
→ The detection and measurement of electric charges
→ Coulomb's law
→ The electric field
→ Electric potential energy
→ Electric potential
→ Capacitance

The Nature of Electric Charges

The ancient Greeks used to rub pieces of amber on wool or fur. The amber was then able to pick up small objects that were not made of metal. The Greek word for amber was *elektron*—hence the term *electric*. The pieces of amber would retain their attractive property for some time, so the effect appears to have been **static**. The amber acted differently from magnetic ores (lodestones), naturally occurring rocks that attract only metallic objects.

In modern times, hard rubber, such as ebonite, is used with cloth or fur to dramatically demonstrate the properties of electrostatic force. If you rub an ebonite rod with cloth (charging by friction) and then bring it near a small cork sphere painted silver (called a "pith ball") and suspended on a thin thread, the pith ball will be attracted to the rod. When the ball and rod touch each other, the pith ball will be repelled. If a glass rod that has been rubbed with silk is then brought near the pith ball, the ball will be attracted to the rod. If you touch the pith ball with your hand, the pith ball will return to its normal state. This is illustrated in Figure 2.1.

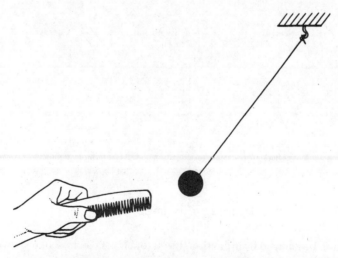

Figure 2.1 Electrostatic Attraction

In the nineteenth century, chemical experiments to explain these effects showed the presence of molecules called **ions** in solution. These ions possessed similar affinities for certain objects, such as carbon or metals, placed in the solution. These objects are called **electrodes**. The experiments confirmed the existence of two types of ions, **positive** and **negative**. The effects they produce are similar to the two types of effects produced when ebonite and glass are rubbed. Even though both substances attract small objects, these objects become **charged** oppositely when rubbed, as indicated by the behavior of the pith ball. Further chemical experiments coupled with an atomic theory demonstrated that in solids it is the negative charges that are transferred. Additional experiments by Michael Faraday in England during the first half of the nineteenth century suggested the existence of a single fundamental carrier of negative **electric charge**, which was later named the **electron**. The corresponding carrier of positive charge was termed the **proton**.

The Detection and Measurement of Electric Charges

When ebonite is rubbed with cloth, only the part of the rod in contact with the cloth becomes charged. The charge remains localized for some time (hence the name *static*). For this reason, among others, rubber, along with plastic and glass, is called an **insulator**. A metal rod held in your hand cannot be charged statically for two reasons. First, metals are **conductors**; that is, they allow electric charges to flow through them. Second, your body is a conductor, and any charges placed in the metal rod are conducted out through you (and into the earth). This effect is called **grounding**. The silver-coated pith balls mentioned earlier can become statically charged because they are suspended by thread, which is an insulator. They can be used to detect the presence and sign of an electric charge, but they are not very helpful in obtaining a qualitative measurement of the magnitude of charge they possess.

An instrument that is often used for qualitative measurement is the **electroscope**. One form of electroscope consists of two "leaves" made of gold foil (Figure 2.2a). The leaves are vertical when the electroscope is uncharged. As a negatively charged rod is brought near, the leaves diverge. If we recall the hypothesis that only negative charges move in solids, we can understand that the electrons in the knob of the electroscope are repelled down to the leaves through the conducting stem. The knob becomes positively charged, as can be verified with a charged pith ball, as long as the rod is near but not touching (Figure 2.2b). If you take the rod away, the leaves will collapse as the electroscope is still neutral overall.

Upon contact, electrons are directly transferred to the knob, stem, and leaves. The whole electroscope then becomes negatively charged (Figure 2.2c). The extent to which the leaves are spread apart is an indication of how much charge is present (but only qualitatively). If you touch the electroscope, you will ground it, and the leaves will collapse together.

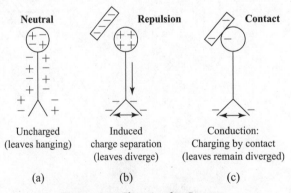

Figure 2.2 Charging by Contact

The electroscope can also be charged by **induction**. If you touch the electroscope shown in Figure 2.2b with your finger when the electroscope is brought near (see Figure 2.3a), the repelled electrons will be forced out into your body. If you remove your finger, keeping the rod near, the electroscope will be left with an overall positive charge by induction (Figure 2.3b).

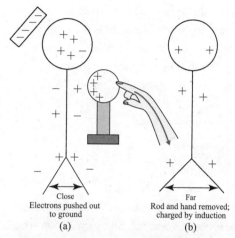

Figure 2.3 Charging by Induction

Finally, we can state that electric charges, in any distribution, obey a conservation law. When we transfer charge, we always maintain a balanced accounting. Suppose we have two charged metal spheres. Sphere *A* has +5 elementary charges, and sphere *B* has +1 elementary charge (thus both are positively charged). The two spheres are brought into contact. Which way will charges flow? Excess charges are always spaced out as far apart as possible since they repel each other. When charges are allowed to flow, electrons do the moving even if the net charge on both objects is positive. Note that the vast majority of charge is not moving! If the two spheres are of equal size, they will each have a +3 charge after enough electrons move from the +1 sphere to the +5 sphere in this case. If one object is bigger than the other, the larger object will wind up with more of the net excess charge after those excess charges have all spread out as evenly as possible around the outer surfaces of the combined object.

Coulomb's Law

From the first two sections of this chapter, we can conclude that like charges repel (see Figure 2.4) each other while unlike charges attract. The electrostatic force between two charged objects can act through space and even a vacuum. This property makes electrostatic force similar to the force of gravity.

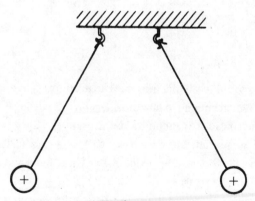

Figure 2.4 Similar Charges Repel

In the SI system of units, charge is measured in **coulombs** (C), and one elementary charge is designated as *e* and has a magnitude of 1.6×10^{-19} C. In the late eighteenth century, the nature of the electrostatic force was studied by French scientist Charles Coulomb. He discovered that the force between two point charges

Think About It

Coulomb's law is an *inverse square* law that is very similar to Newton's law of gravitation. However, the electrostatic force can be neutralized, while gravity is always present. Additionally, with the electrostatic force, there can be repulsion as well as attraction.

(designated as q_1 and q_2), separated by a distance r, experienced a mutual force along a line connecting the charges that varied directly as the product of the charges and inversely as the square of the distance between them. This law, known as **Coulomb's law**, is, like the law of gravity, an inverse square law acting on matter at a distance. Mathematically, Coulomb's law can be written as

> Coulomb's constant k can be written more formally as
>
> $$k = \frac{1}{4\pi\varepsilon_0}$$
>
> where ε_0 is the electrical permittivity of free space (that is, a vacuum).

$$F = \frac{kq_1 q_2}{r^2}$$

The constant k has the value $8.9875 \times 10^9 \approx 9 \times 10^9 \, \mathrm{N \cdot m^2/C^2}$.

Coulomb's law, like Newton's law of universal gravitation, is a vector equation. The direction of the force is along a radial vector connecting the two point sources. It should be noted that Coulomb's law applies only to one pair of point sources (or to sources that can be treated as point sources, such as charged spheres). If we have a distribution of point charges, the net force on one charge is the vector sum of all the other electrostatic forces. This aspect of force addition is sometimes termed **superposition**.

> **NOTE**
>
> On the AP Physics 2 exam, you will only encounter superposition problems for up to four individual charges. (This is true for force, energy, and field calculations.)

> **Sample Problem 1**

Calculate the static electric force between a $+6.0 \times 10^{-6}$ C charge and a -3.0×10^{-6} C charge separated by 0.1 m. Is this an attractive or repulsive force?

> **Solution**

We use Coulomb's law:

$$F = \frac{kq_1 q_2}{r^2}$$

$$F = \frac{(9 \times 10^9 \, \mathrm{N \cdot m^2/C^2})(+6.0 \times 10^{-6} \, \mathrm{C})(-3.0 \times 10^{-6} \, \mathrm{C})}{(0.1 \, \mathrm{m})^2} = -16.2 \, \mathrm{N}$$

The negative sign (opposite charges) indicates that the force is attractive.

The Electric Field

Another way to consider the force between two point charges is to recall that the force can act through free space. If a charged sphere has charge $+Q$, and a small test charge (small enough to have no effect on the existing electric field) $+q$ is brought near it, the test charge will be repelled according to Coulomb's law. Everywhere, the test charge will be repelled along a radial vector out from charge $+Q$. We can state that, even if charge $+Q$ is too small to be visible, the influence of the electrostatic force can be observed and measured (since charges have mass and $\vec{E} = m\vec{a}$ as usual).

In this way, charge $+Q$ is said to set up an electric field, which pervades the space surrounding the charge and produces a force on any other charge that enters the field (just as a gravitational field does). The strength of the electric field, $\vec{E}$, is defined to be the measure of the force per unit charge experienced at a particular location. In other words,

$$\vec{E} = \frac{\vec{F}}{q}$$

and the units are newtons per coulomb (N/C). The electric field strength $\vec{E}$ is a vector quantity since it is derived from the force (a vector) and the charge (a scalar). The direction of the electric field is defined to be the direction of the force on a positive charge at that location.

An illustration of this situation is given in Figure 2.5. Charge $+Q$ is represented by a charged sphere drawn as a circle. A small positive test charge, $+q$, is brought near and repelled along a radial vector $\vec{r}$ drawn outward from charge $+Q$. In fact, anywhere in the vicinity of charge $+Q$, the test charge will be repelled along an outward radial vector. We therefore draw these **force field lines** as radial vectors coming out from charge $+Q$.

By convention, we always consider the test charge to be positive. Since the force varies according to Coulomb's law, the electric field strength $\vec{E}$ will also vary, depending on the location of the test charge.

> **TIP**
>
> **Compare this formula for the electric field to the formula for the gravitational field strength $\vec{g} = \vec{F}/m$.**

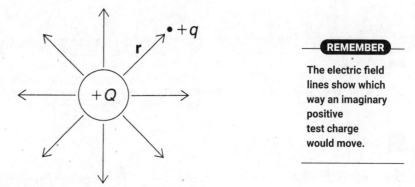

> **REMEMBER**
>
> **The electric field lines show which way an imaginary positive test charge would move.**

Figure 2.5 Drawing Electric Field Lines

Combining the formula for Coulomb's law and our definition of electric field, the electric field around a point (of spherical distribution) of charge is

$$|\vec{E}| = \frac{|q|}{4\pi\varepsilon_0 r^2}$$

where q is the source charge creating the electric field and r is the radial distance from the center of that charge.

We can also interpret field strength by observing the density of field lines per square meter. With point charges, the radial nature of their construction causes the field lines to converge near the surface of the charge, Q, indicating a relative increase in field strength. Later in this chapter, we will encounter a configuration in which the field strength remains constant throughout.

There are several other configurations of electric field lines that we can consider. In Figure 2.6, the field between two point charges, with different arrangements of signs, is illustrated. Notice for a conductor that the field lines are always perpendicular to the "surface" of the source and that they never cross each other. This fact can be extended to the idea that the electric field inside a hollow conducting sphere is zero and all of the charges reside on the surface of the sphere.

The arrows on the field lines always indicate the direction in which a positive test charge would move.

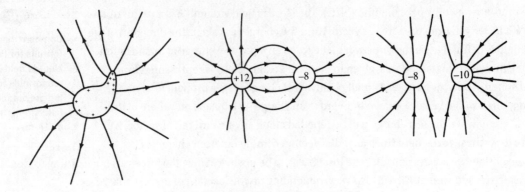

Figure 2.6 Examples of Electric Fields

> **Sample Problem 2**

What is the force acting on an electron placed in an external electric field $E = 100$ N/C?

✓ **Solution**

We know that $\vec{E} = \vec{F}/q$.

Thus,

$$\vec{F} = \vec{E}/q = (100 \text{ N/C})(-1.6 \times 10^{-19} \text{ C}) = -1.6 \times 10^{-17} \text{ N}$$

Electric Potential Energy

Electrical forces are conservative forces in the same sense that the force of gravity is conservative: The work done by these forces is path independent. In other words, when calculating the work done between two fixed points by any conservative force, one gets the same answer regardless of the details of the particular path traveled. Recall that the potential energy of a conservative force is defined in terms of the work done by that force. Work is equal to the magnitude of force times displacement in the direction of the force. So, for a charge q being displaced in an electric field E over a distance Δr:

$$W = Eq\Delta r$$

In the special case of Coulomb's law, this gives us an expression for the potential energy (U) contained in the relationship between the two interacting charges:

$$U_{\text{elec}} = \frac{q_1 q_2}{4\pi\varepsilon_0 r} = k\frac{q_1 q_2}{r}$$

Many points can have the same electric potential energy. Consider the original electric field diagram in Figure 2.5. If test charge $+q$ is a distance r from source charge $+Q$, it will experience a certain radial force whose magnitude is given by Coulomb's law. At any position around the source, at a fixed distance r, we can observe that the test charge will experience the same force. The localized potential energy will be the same as well. The set of all such positions defines a circle of radius r (actually a sphere in three dimensions) called an **equipotential surface**.

Electric Potential

Figure 2.7 shows a series of equipotential "curves" of varying radii. Since the electric field is conservative, work is done by or against the field only when a charge is moved from one equipotential surface to another. To better understand this effect, we define a quantity called the **electric potential**, which is a measure of the magnitude of electric potential energy per unit charge at a particular location in the field. Thus, the work done per unit charge, in moving from equipotential surface A to equipotential surface B, is a measure of what is called the **potential difference** or **voltage**. Equipotential lines are sometimes referred to as isolines of electric potential. Note that there can be no component of the electric field in the direction of these isolines.

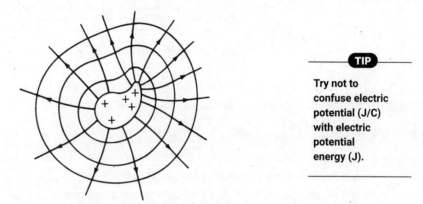

TIP

Try not to confuse electric potential (J/C) with electric potential energy (J).

Figure 2.7 Equipotentials in an Electric Field

We thus define electric potential difference (V or ΔV) to be the potential difference between two points ($V_B - V_A$). Note that voltage or emf (electromotive force) are sometimes used as synonyms for "electric potential difference." Since electric potential difference is the energy per charge, the units of potential difference are joules/coulomb or volts:

$$\Delta V = \frac{\Delta U_{\text{elec}}}{q}$$

Units of electric field potential = volt = V = J/C = N · m/C

Sometimes it is convenient to speak of the potential at a point. This implies that a zero-volt point has already been chosen and all the other electric potential differences are relative to that point. Frequently, zero voltage is taken at infinity. At other times, it is taken to be at the closest grounding point or the negative terminal of a battery in a simple circuit.

If there exists a constant, uniform electric field (see below), then rearranging the definition of electric potential difference shown previously yields the following relationship between electric field (a proxy for electric force) and potential difference (a proxy for electric potential energy):

$$|\vec{E}| = \left|\frac{\Delta V}{\Delta r}\right|$$

Units of electric field = N/C = V/m

Therefore, if we have a uniform electric field, the electric potential difference between two points is simply the electric field times the distance between the two points in the direction of the field. One way to produce such a uniform electric field is between two charged parallel plates that are separated by a distance d. If each side is oppositely charged (see Figure 2.8), the electric field will be uniform. Therefore, a test charge within this field would feel the same force at any point in between the plates. Compare this to the gravitational field g near Earth's surface. A test mass in the location experiences the same force mg anywhere in this area. Note in Figure 2.8 that the electric field loses its uniformity near the edges of the plates. This phenomenon is known as the **fringe effect**. It is due to the fact that the charge distribution is no longer uniform at the edges.

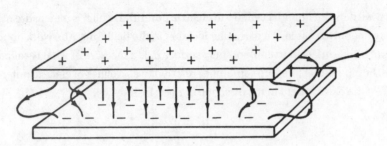

Figure 2.8 Constant Electric Field Between Plates

Recall that work done by a conservative force is the negative of the change in potential energy. By combining our definition of electric potential difference with our work relationship to potential energy:

$$W = q\Delta V = -\Delta U$$

If the moving charge is positive, a positive ΔV results in negative work (the charge is losing energy). However, if the moving charge is negative, a positive ΔV results in positive work (the charge is gaining energy). Just as positive and negative charges will experience opposite forces when placed into the same electric field, so, too, will they lose energy by going in opposite directions. Generally speaking, positive charges move from high electric potential differences to low electric potential differences, whereas negative charges move from low electric potential differences to high ones. Going back to our gravitational analogy, negative mass (if it existed) would experience an upward force. By moving up, it would lower its gravitational potential energy by raising its gravitational potential. In this analogy, height is electric potential difference while the ground is the negative plate. The negative mass is certainly gaining height (greater electric potential difference) by responding to the upward force. Since its mass is negative, its gravitational potential energy is decreasing.

Another convenient unit of electrical energy is the **electron volt** (eV). By definition, if one elementary charge experiences a potential difference of 1 volt, its energy (or the work done to transfer the charge) is equal to 1 electron volt. Consequently, we can state that 1 electron volt $= 1.6 \times 10^{-19}$ joule.

> **Sample Problem 3**

Given that the charge on a proton is $+1.6 \times 10^{-19}$ C,

(a)   Calculate the electric potential at a point 2.12×10^{-10} m from the proton.

(b)   If an electron (and no other charges) is placed at that point, what will be its electric potential energy? Assume that the potential energy at infinity is equal to zero.

✓ **Solution**

(a)   For the proton, we use the work done from Coulomb's law to obtain the expression for the voltage around a point charge Q:

$$V = \frac{kQ}{r}$$

$$V = \frac{(9 \times 10^9 \, \text{N} \cdot \text{m}^2/\text{C}^2)(1.6 \times 10^{-19} \, \text{C})}{2.12 \times 10^{-10} \, \text{m}^2} = 6.79 \, \text{V}$$

(b)   For the electron, the potential energy is:

$$U = qV = (-1.6 \times 10^{-19} \, \text{C})(6.79 \, \text{V}) = -1.09 \times 10^{-18} \, \text{J}$$

Although abstract in nature, electrical force and energy fields can often be more easily understood by analogy with the more familiar fields for gravity. The underlying algebraic similarity is shown in Table 2.1. Figure 2.9 shows the structural similarities between the electrical and gravitational fields.

Table 2.1 Similarities in Uniform Gravitational and Electrical Formulas

	Gravitational	Electrical
Forces	mg	qE
Potential Energy Differences	$mg\Delta h$	$qE\Delta x$

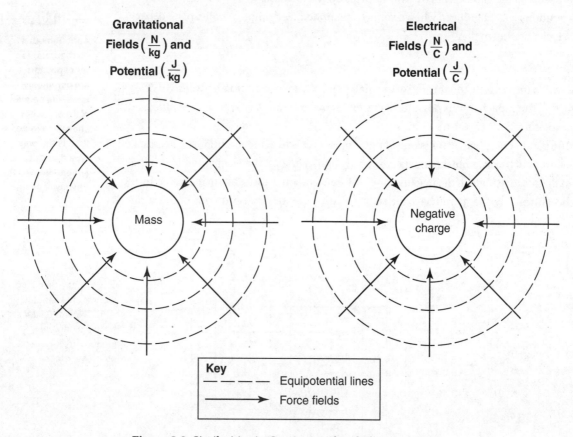

Gravitational
Fields $\left(\dfrac{N}{kg}\right)$ and
Potential $\left(\dfrac{J}{kg}\right)$

Electrical
Fields $\left(\dfrac{N}{C}\right)$ and
Potential $\left(\dfrac{J}{C}\right)$

Mass

Negative charge

Key
— — — Equipotential lines
———▶ Force fields

Figure 2.9 Similarities in Gravitational and Electrical Fields

Capacitance

When there are two charged parallel metal plates in a configuration called a **capacitor**, there is a proportional relationship between the potential difference and the total charge. As the external voltage supply pushes a charge onto one side of the capacitor, that side will become charged. This charge will push the same amount and sign of charge out of the capacitor on the opposite side. Thus, the capacitor gains an equal but opposite amount of charge on each side, maintaining an overall neutral state. As the charges accumulate, it becomes increasingly difficult to add more charge, and the charging of the capacitor exponentially levels off. Because of this, the voltage across and the current through the capacitor also level off exponentially. As the charge and voltage increase, the current will decrease. The capacitance, C, of a capacitor is defined to be the ratio of the charge on either conductor and the potential difference between the conductors:

$$C = \frac{Q}{V}$$

This is a positive scalar quantity that is essentially a measure of the capacitor's ability to store charge. The reason is that, for a given capacitor, the ratio Q/V is a constant since potential difference increases as the charge on either capacitor increases.

The SI unit of capacitance is the **farad** (F). The types of capacitors typically used in electronic devices have capacitances that range from microfarads to picofarads. The capacitance of a given object is a property of the materials used and the geometry of the two surfaces upon which the two oppositely charged distributions will be induced when a voltage is supplied. For simple parallel plate capacitors, it can be shown that

$$C = Q/V = \varepsilon_0 A/d$$

where A is the cross-sectional area of the plates and d is the separation between them. Note that the capacitance is not affected by either the amount of charge or the voltage but, instead, is fixed by its construction.

The parallel plate capacitor is one of the simplest ways to achieve a uniform electric field. In Figure 2.10 below, note the regularly spaced parallel lines of the electric field pointing from the positive plate to the negative one. The equipotential surfaces are indicated by the vertical lines. Any charge will have the same electric potential energy on one of the surfaces.

> **NOTE**
>
> Although other configurations for capacitors are commonly used, the parallel plate capacitor shown in Figure 2.10 is the only type that will appear on the AP Physics 2 exam.

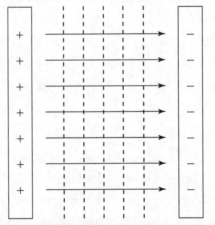

Figure 2.10 Equipotential Lines in a Constant Electric Field

> Note that the fields are uniform only near the center. Near the edges, the fields become nonuniform, known as the fringe effect.

Assuming the field is truly uniform, as shown, the electric field between the two plates will be constant and determined uniquely by the charge (Q) on each plate and its surface area (A):

$$E = \frac{Q}{\varepsilon_0 A}$$

Many capacitors are manufactured with insulating material inserted in the space between the plates to increase the maximum charge that can be stored (and thus the capacitance of the device itself). In this case, the insulating material is characterized by a dielectric constant, κ, which modifies the permittivity of free space such that all of the above equations for capacitors remain the same with the substitution:

$$\varepsilon_0 \rightarrow \kappa \varepsilon_0$$

The energy stored in a capacitor can be studied by recognizing that the definition of capacitance can be written as $Q = CV$. This direct relationship between plate charge and voltage will produce a diagonal straight line if we plot a graph of charge versus voltage (see Figure 2.11).

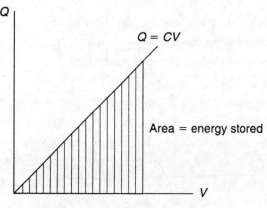

Figure 2.11 Energy in Capacitors

Notice that the units of QV are joules (the units of work). The area under the diagonal line is equal to the amount of energy stored in the capacitor. This energy is given by $E = (½)CV^2$.

> ### Sample Problem 4

A point charge $+Q$ is fixed at the origin of a coordinate system. A second charge $-q$ is uniformly distributed over a spherical mass m and orbits around the origin at a radius r and with a period T. Derive an expression for the radius.

✓ Solution

If we neglect gravity and other external forces, the orbit is due to the electrostatic force between the charges. This sets up a centripetal force for the mass m.

Thus, we can write

$$F_c = m\frac{v^2}{r}$$

$$k\frac{Qq}{r^2} = m\frac{\left(\frac{2\pi r}{T}\right)^2}{r}$$

Simplifying and solving for r gives

$$r = \left(\frac{kQqT^2}{4\pi^2 m}\right)^{\frac{1}{3}}$$

SUMMARY

- There are two kinds of electric charges: positive and negative.
- Electrons are the fundamental carriers of negative charge.
- Protons carry a positive charge equal in magnitude to that of the electron.
- Like charges repel, while unlike charges attract.
- An electroscope can be used to detect the presence of static charges.
- Objects become charged through the transfer of electrons.
- Coulomb's law describes the nature of the force between two static charges. It states that the force of either attraction ($-$) or repulsion ($+$) is proportional to the product of the charges and inversely proportional to the square of the distance between them. This is similar to Newton's law of gravitation.
- The electric potential difference is defined to be equal to the work done per unit charge (in units of joules per coulomb or volts).
- A capacitor is a device that can store charge and energy. The capacitance of a capacitor is defined as the amount of charge stored per unit volt ($C = Q/V$).
- Electric field lines indicate the direction of electric force on positive charges. They come out of positive source charges and go into negative source charges. They never cross, and their relative spacing indicates the strength of the field.
- Equipotential lines indicate regions of constant electric potential. The lines (surfaces, actually) intersect the electric field lines at right angles and have their highest values near positive source charges and their lowest values near negative source charges.

Problem-Solving Strategies for Electrostatics

The preceding sample problem reminds us that we are dealing with vector quantities when the electrostatic force is involved. Drawing a sketch of the situation and using our techniques from vector constructions and algebra were most effective in solving the problem.

Be sure to keep track of units. If a problem involves potential or capacitance, remember to maintain the standard SI system of units, which are summarized in the Appendices.

Remember that electric field lines are drawn as though they followed a "positive" test charge and that Coulomb's law applies only to point charges. The law of superposition allows you to combine the effects of many forces (or fields) using vector addition. Additionally, symmetry may allow you to work with a reduced situation that can be simply doubled or quadrupled. In this aspect, a careful look at the geometry of the situation is called for.

If the electric field is uniform, the charges in motion will experience a uniform acceleration that will allow you to use Newtonian kinematics to analyze the motion. Electrons in oscilloscopes are an example of this kind of application.

Potential and potential difference (voltage) are scalar quantities usually measured relative to the point at infinity in which the localized potential is taken to be zero. In a practical sense, it is the change in potential that is electrically significant, not the potential itself.

Practice Exercises

1. An insulated metal sphere, A, is charged to a value of $+Q$ elementary charges. It is then touched and separated from an identical but neutral insulated metal sphere, B. This second sphere is then touched to a third identical and neutral insulated metal sphere, C. Finally, spheres A and C are touched and then separated. Which of the following represents the distribution of charge on each sphere, respectively, after the above process has been completed?

    (A) $Q/3, Q/3, Q/3$
    (B) $Q/4, Q/2, Q/4$
    (C) $3Q/8, Q/4, 3Q/8$
    (D) $3Q/8, Q/2, Q/4$

2. A parallel-plate capacitor has a capacitance of C. If the area of the plates is doubled while the separation between the plates is halved, the new capacitance will be

    (A) $4C$.
    (B) $2C$.
    (C) C.
    (D) $C/2$.

3. What is the capacitance of a parallel-plate capacitor made of two aluminum plates, 4 cm in length on a side and separated by 5 mm?

    (A) 2.832×10^{-12} F
    (B) 2.832×10^{-11} F
    (C) 2.832×10^{-10} F
    (D) 2.832×10^{-9} F

4. If 10 J of work are required to move 2 C of charge in a uniform electric field, the potential difference present is equal to

    (A) 20 V.
    (B) 12 V.
    (C) 8 V.
    (D) 5 V.

5. Which of the following diagrams represents the equipotential curves in the region between a positive point charge and a negatively charged parallel plate?

    (A)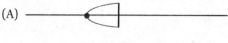

    (B)

    (C)

    (D)

6. An electron is placed between two charged parallel plates as described below. Which of the following statements are true?

    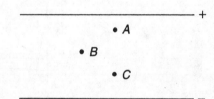

    I.   The electrostatic force is greater at A than at B.

    II. The work done from A to B to C is the same as the work done from A to C.

    III. The electrostatic force is the same at points A and C.

    IV. The electric field strength decreases as the electron is repelled upward.

    (A) I and II
    (B) I and III
    (C) II and III
    (D) II and IV

7. How much kinetic energy is given to a doubly ionized helium atom due to a potential difference of 1,000 V?

    (A) 3.2×10^{-19} eV
    (B) 3.2×10^{-16} eV
    (C) 1,000 eV
    (D) 2,000 eV

8. Which of the following is equivalent to 1 F of capacitance?

    (A) $\dfrac{C^2 \cdot s^2}{kg \cdot m^2}$

    (B) $\dfrac{kg \cdot m^2}{C^2 \cdot s^2}$

    (C) $\dfrac{C}{kg \cdot s}$

    (D) $\dfrac{C \cdot m}{kg^2 \cdot s}$

9. An electron enters a uniform electric field between two parallel plates vertically separated with $\vec{E} = 400$ N/C. The electron enters with an initial velocity of 4×10^7 m/s. The length of each plate is 0.15 m.

    (a) What is the acceleration of the electron?

    (b) How long does the electron travel through the electric field?

    (c) If it is assumed that the electron enters the field level with the top plate, which is negatively charged, what is the vertical displacement of the electron during that time?

    (d) What is the magnitude of the velocity of the electron as it exits the field?

10. A 1 g cork sphere on a string is coated with silver paint. It is in equilibrium, making a 10° angle in a uniform electric field of 100 N/C as shown below. What is the charge on the sphere?

11. An electrostatically charged rod attracts a small, suspended sphere. What conclusions can you make about the charge on the suspended sphere?

12. Why is it safer to be inside a car during a lightning storm?

Answer Explanations

1. **(C)** We use the conservation of charge to determine the answer. Sphere A has $+Q$ charge, and B has zero. When these spheres are touched and separated, each will have one-half of the total charge, which was $+Q$. Thus, each now has $+Q/2$.

   When B is touched with C, which also has zero charge, we again distribute the charge evenly by averaging. Thus, B and C will now each have $+Q/4$, while A still has $+Q/2$.

   Finally, when A and C are touched, we take the average of $+Q/4$ and $+Q/2$, which is $+3Q/8$.

   The final distribution is therefore $3Q/8$, $Q/4$, $3Q/8$.

2. **(A)** The formula for the capacitance of a parallel-plate capacitor is $C = \varepsilon_0 A/d$. Therefore, if the area is doubled and the separation distance is halved, the capacitance increases by four times and is equal to $4C$.

3. **(A)** We use the formula from question 2, but first all measurements must be in meters or square meters. Converting the given lengths yields 4 cm $= 0.04$ m and 5 mm $= 0.005$ m. Then $A = (0.04)(0.04) = 0.0016$ m^2 and $d = 0.005$ m. Using the formula and the value for the permittivity of free space given in the chapter, we get $C = 2.832 \times 10^{-12}$ F.

4. **(D)** Potential difference is the work done per unit charge. Thus

   $$V = \frac{W}{q} = \frac{10}{2} = 5 \text{ V}$$

5. **(C)** The equipotential curves around a point charge are concentric circles. The equipotential curves between two parallel plates are parallel lines (since the electric field is uniform). The combination of these two produces the diagram shown in choice (C).

6. **(C)** The electric field is the same at all points between two parallel plates since it is uniform. Thus, the force on an electron is the same everywhere. Also, the electric field, like gravity, is a conservative field. This means that work done to or by the field is independent of the path taken. Thus, both statements II and III are true.

7. **(D)** One electron volt is defined to be the energy given to one elementary charge through 1 V of potential difference. A doubly ionized helium atom has $+2$ elementary charges. Since the potential difference is 1,000 V, the kinetic energy is 2,000 eV. When using electron volt units, we eliminate the need for the small numbers associated with actual charges of particles.

8. **(A)** One farad is defined to be 1 C/V. One volt is 1 J/C. One joule is 1 kg $\cdot$ m^2/s^2. The combination results in

   $$\frac{C^2 \cdot s^2}{kg \cdot m^2}$$

9. (a) Since the electron is in a uniform electric field, the kinematics of the electron follows the standard kinematics of mechanics. Since

$$\vec{F} = m\vec{a} \text{ and } \vec{F} = -e\vec{E}, \vec{a} = -e\vec{E}/m$$

(the negative sign is needed since an electron is negatively charged). Using the information given in the problem and the values for the charge and mass of the electron from the Appendices, we get

$$a = \frac{-(1.6 \times 10^{-19})(400)}{9.1 \times 10^{-31}} = -7.03 \times 10^{13} \text{ m/s}^2$$

(b) The electron is subject to a uniform acceleration in the downward direction. The path will therefore be a parabola, and the electron will be a projectile in the field. The initial horizontal velocity remains constant, and therefore the time is given by the expression

$$t = \frac{\ell}{v_0} = \frac{0.15}{4 \times 10^7} = 3.75 \times 10^{-9} \text{ s}$$

(c) If we take the initial entry level as zero in the vertical direction, the electron has initially zero velocity in that direction. Thus, the vertical displacement is given by $y = (1/2)at^2$. Using the information from parts (a) and (b), we get

$$y = -(1/2)(7.03 \times 10^{13})(3.75 \times 10^{-9})^2 = -4.94 \times 10^{-4} \text{ m}$$

(d) The magnitude of the velocity as the electron exits the field, after time t has elapsed, is given by the vector nature of two-dimensional motion. The horizontal velocity is the same throughout the encounter and is equal to 4×10^7 m/s. The vertical velocity is given by $v_y = -at$ since there is no initial vertical velocity. Using the information from the previous parts, we find that

$$v_y = -(7.03 \times 10^{13})(3.75 \times 10^{-9}) = -2.6 \times 10^5 \text{ m/s}$$

Now the magnitude of the velocity is given by the Pythagorean theorem since the velocity is the vector sum of these two component velocities:

$$v = \sqrt{(4 \times 10^7)^2 + (2.6 \times 10^5)^2} = 40,000,845 \text{ m/s} \approx 4 \times 10^7 \text{ m/s}$$

10. The result is not a significant change in the magnitude of the initial velocity. There is of course a significant change in direction. The angle of this change can be determined by the tangent function if desired.

If the sphere is in equilibrium, the vector sum of all forces acting on it must equal zero. There are several forces involved. Mechanically, gravity acts to create a tension in the string. This tension has an upward component that directly balances the force of gravity (mg) acting on the sphere. A second horizontal component acts to the left and counters the horizontal electrical force established by the field and the charge on the sphere. A free-body diagram looks like this:

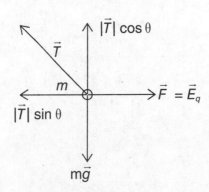

From the diagram, we see that $T\cos\theta = mg$, where $\theta = 10°$ and $m = 1\,\text{g} = 0.001\,\text{kg}$. Thus, $T = 0.00995\,\text{N}$, using our known value for the acceleration due to gravity, g. Now, in the horizontal direction, $T\sin\theta = E_q$ and $E = 100\,\text{N/C}$. Using all known values given and derived, we arrive at

$$q = 1.75 \times 10^{-5}\,\text{C}$$

11. The only conclusion you can make is that the sphere is either "neutral" or oppositely charged compared with the rod.

12. If your car is hit by lightning, the charges are distributed around the outside and then dissipated away. In a sense, this acts like an electrostatic shield.

3

Electric Circuits

Current and Electricity

In Chapter 2, we observed that if two points have a potential difference between them and they are connected with a conductor, negative charges will flow from a higher concentration to a lower one. This aspect of charge flow is very similar to the flow of water in a pipe due to a pressure difference.

Moving electric charges are referred to as electric **current**, which measures the amount of charge passing a given point every second. The units of measurement are coulombs per second (C/s). These are defined as **amperes** or "amps." Algebraically, we designate current by the capital letter I and state that

$$I = \frac{\Delta Q}{\Delta t}$$

In electricity, it is the battery that supplies the potential difference needed to maintain a continuous flow of charge. In the nineteenth century, physicists thought that this potential difference was an electric force, called the **electromotive force** (emf), that pushed an electric fluid through a conductor. Today, we know that an emf is not a force but, rather, a potential difference measured in volts. Do not be confused by the designation "emf" in the course of reviewing or of solving problems!

From chemistry, recall that a battery uses the action of acids and bases on different metals to free electrons and maintain a potential difference. In the process, two terminals, designated positive and negative, are created. When a conducting wire is attached and looped around to the other end, a complete circle of wire (a **circuit**) is produced, allowing for the continuous flow of charge. The battery acts like an elevator, raising electrons from the positive side up to the negative side using chemical reactions (see Figure 3.1). These electrons can then do work by transforming their electric potential energy into other forms of energy. This work is the electricity with which we have become so familiar in our modern world.

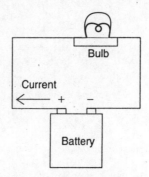

Figure 3.1 A Simple Circuit

The diagram in Figure 3.1 shows a simple electric circuit. The direction of the conventional current, like electric fields, is from the positive terminal. To maintain a universal acceptance of concepts and ideas, schematic representations for electrical devices were developed and accepted by physicists and electricians worldwide. These schematics are used when drawing or diagramming an electric circuit, and it is important that you are able to interpret and draw them to fully understand this topic. In Figure 3.2, schematics for some of the most frequently encountered electrical devices are presented.

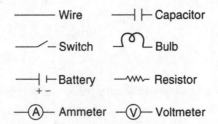

Figure 3.2 Electrical Schematic Diagrams

The simple circuit shown in Figure 3.1 can now be diagrammed schematically as in Figure 3.3.

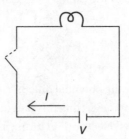

Figure 3.3 A Simple Circuit Schematic

A switch has been added to the schematic; of course, the charge would not flow unless the switch was closed. An open switch stops the flow by breaking the circuit.

In Figure 3.2, three schematics appear that we have not yet discussed. A **resistor** is a device whose function is to use up voltage. We will investigate resistors in more detail in the next section.

An **ammeter** is a device that measures the current. You can locate the water meter in your house or apartment building and notice that it is placed within the flow line. The reason is that the meter must measure the flow of water per second through a given point. An ammeter is placed within an electric circuit in much the same way. This is referred to as a "series connection," and it maintains the singular nature of the circuit. In practical terms, you can imagine cutting a wire in Figure 3.3 and hooking up the bare leads to the two terminals of the ammeter. Ammeters have very low resistance.

A **voltmeter** is a device that measures the potential difference, or voltage, between two points. Unlike an ammeter, the voltmeter cannot be placed within the circuit since it will effectively be connected to only one point. A voltmeter is therefore attached in a "parallel connection," creating a second circuit through which only a small amount of current flows to operate the voltmeter. In Figure 3.4, the simple circuit is redrawn with the ammeter and voltmeter placed. Voltmeters have very high resistance.

A volt is the unit of electric potential; 1 volt = 1 joule/coulomb. A volt represents the amount of energy per charge relative to either a predetermined "ground" ($V = 0$) or between two points. Voltage, potential, potential difference, electromotive force (emf), voltage drop, and electric potential are all terms used for this important concept in electricity (recall that gravitational potential energy works similarly). *Potential difference* is the preferred term on the AP Physics 2 exam.

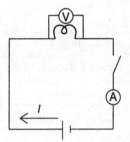

Figure 3.4 A Simple Circuit with Voltmeter and Ammeter

For simple circuits, there will be no observable difference in readings if the ammeter and voltmeter are moved to different locations. However, there is a slight difference in the emf across the terminals of the battery when the switch is closed versus when it is open. The first emf reflects the work done by the battery not only to move charge through the circuit but also to move charge across the terminals of the battery. This is sometimes referred to as the **terminal emf** or **internal emf**.

When a battery dies, it doesn't run out of volts or charge. It runs out of energy. The battery's voltage rating is a measure of how much energy it will give up to each coulomb of charge that passes through it. The battery will continue to give up its energy until it runs out of energy. When you recharge a battery, you are replenishing its energy!

Electrical Resistance

In Figure 3.4, a simple electric circuit is illustrated with measuring devices for voltage and current. If a light bulb is left on for a long time, two observations can be made. First, the bulb gets hot because of the action of the electricity in the filament. The light produced by the bulb is caused by the heat of the filament. Second, the current in the ammeter will begin to decrease.

These two observations are linked to the idea of electrical resistance. The interaction of flowing electrons and the molecules of a wire (or bulb filament) creates an electrical **resistance**. This resistance is temperature dependent since, from the kinetic theory, an increase in temperature will increase the molecular activity and therefore interfere to a greater extent with the flow of current. Electric potential energy is being converted to thermal energy.

In the case of a light bulb, it is this resistance that is desired in order for the bulb to do its job. Resistance along a wire or in a battery, however, is unwanted and must be minimized. In more complicated circuits, a change in current flow is required to protect devices, and so special resistors are manufactured that are small enough to easily fit into a circuit. While resistance is the opposite of conductance, we do not want to use insulators as resistors since insulators will stop the flow altogether. Therefore, a range of materials, indexed by "resistivity," is catalogued in electrical handbooks to assist scientists and electricians in choosing the proper resistor for a given situation.

If the temperature can be maintained at a constant level, a simple relationship between the voltage and the current in a circuit is revealed: as the voltage is increased, a greater flow of current is observed. This direct relationship was investigated theoretically by a German physicist named Georg Ohm and is called **Ohm's law** (see Figure 3.5).

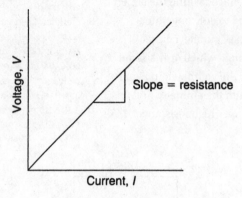

Figure 3.5 Ohm's Law

Ohm's law states that in a circuit at constant temperature the ratio of the voltage to the current remains constant. The slope of the line in Figure 3.5 represents the resistance of the circuit, which is measured in units of ohms (Ω). Algebraically, we can write Ohm's law as

$$R = \frac{\Delta V}{I} \quad \text{or} \quad \Delta V = IR$$

Not all conductors obey Ohm's law. If an object does not have a linear relationship between voltage and current, it is known as "non-Ohmic." Light bulbs, semiconductors, and liquid conductors, for example, do not.

Other factors also affect the resistance of a conductor. We have already discussed the effects of temperature and material type (resistivity). Resistivity is designated by the Greek letter ρ (rho). In a wire, electrons try to move through, interacting with the molecules that make up the wire. If the wire has a small cross-sectional area, the chances of interacting with a bound molecule increase and thus the resistance of the wire increases. If the length of the wire is increased, the greater duration of interaction time will also increase the resistance of the wire. In summary, these resistance factors all contribute to the overall resistance of the circuit.

Algebraically, we may write these relationships (at constant temperature) in the following form:

$$R = \frac{\rho L}{A}$$

where L is the length in meters and A is the cross-sectional area in units of m^2. The material-specific resistivity (ρ) is measured in units of ohm $\cdot$ meters ($\Omega \cdot m$) and is usually rated at 20 degrees Celsius.

In Table 3.1, the resistivities of various materials are presented. Since the ohm is a standard (SI) unit, be sure that all lengths are in meters and areas are in square meters.

Table 3.1 Resistivities of Selected Materials at 20°C*

Substance	Resistivity, ρ ($\Omega \cdot$ m)
Air	$\sim 2 \times 10^{14}$
Aluminum	2.83×10^{-8}
Carbon	3.5×10^{-5}
Copper	1.69×10^{-8}
Glass	$\sim 1 \times 10^{13}$
Gold	2.44×10^{-8}
Quartz	7×10^{17}
Silicon	6.4×10^{2}
Silver	1.59×10^{-8}
Tungsten	5.33×10^{-8}
Wood	$\sim 1 \times 10^{15}$

*Note: Just by looking at the order of magnitude of these resistivities, it is easy to pick out good materials for conductors and insulators.

> Sample Problem 1

Copper wire is being used in a circuit. The wire is 1.2 m long and has a cross-sectional area of 1.2×10^{-8} m² at a constant temperature of 20°C.

(a)   Calculate the resistance of the wire.

(b)   If the wire is connected to a 10 V battery, what current will flow through it?

✓ Solution

(a)   We use

$$R = \rho L / A \text{ where, in this case, } \rho \text{ is the resistivity at 20°C}$$

$$R = \frac{(1.69 \times 10^{-8} \, \Omega \cdot \text{m})(1.2 \, \text{m})}{1.2 \times 10^{-8} \, \text{m}^2} = 1.69 \, \Omega$$

(b)   Now, we use Ohm's law

$$I = \frac{\Delta V}{R} = \frac{10 \, \text{V}}{1.69 \, \Omega} = 5.9 \, \text{A}$$

Electric Power and Energy

Electrical energy can be used to produce light and heat. Electricity can do work to turn a motor. Having measured the voltage and current in a circuit, we can determine the amount of power and energy being produced in the following way. The units of voltage (potential difference) are joules per coulomb (J/C) and are a measure of the energy supplied to each coulomb of charge flowing in the circuit. The current, I, measures the total number of coulombs per second flowing at any given time. The product of the voltage (V) and the current (I) is therefore a measure of the total power produced since the units will be joules per second (watts):

Unit analysis

$$\left(\frac{J}{C}\right) \times \left(\frac{C}{s}\right) = \frac{J}{s} = \text{watt}$$

Derivation

$$P = \frac{\text{work}}{\text{time}}$$

$$= \frac{q\Delta V}{t}$$

$$= \left(\frac{q}{t}\right)\Delta V$$

$$= I\Delta V$$

$$P = \Delta VI$$

The electrical energy expended in a given time t (in seconds) is simply the product of the power and the time:

$$\text{Energy} = Pt = \Delta VIt$$

> ## Sample Problem 2

How much energy is used by a 10 Ω resistor connected to a 24 V battery for 30 minutes?

✓ Solution

We know that the energy is given by $E = \Delta VIt = (\Delta V^2/R)t$, where t must be in seconds.

Thus,

$$E = \frac{(24\text{ V})^2(1{,}800\text{ s})}{10\text{ }\Omega} = 103{,}680\text{ J}$$

TIP

Many problems in physics have an assumption of constant voltage. For example, any appliance that is deriving its electrical power from a wall plug has a set voltage (120 V in the United States). Any appliance being run on batteries will also have the fixed voltage of the battery at all times. So, although in general the voltages may vary, oftentimes the voltage can be assumed to be fixed in a given problem.

Kirchhoff's Rules

Two very important conservation laws underlie all of the circuit analysis that this chapter is about to explore: conservation of charge and conservation of energy. Gustav Kirchhoff first applied these to circuits and gave us the following rules for circuits:

1. The Junction Rule: The total current coming into a junction must equal the total current leaving the junction. (Charge must be conserved.)
2. The Loop Rule: The total voltage drops and gains must total to zero as you travel around any closed loop of a circuit. (Energy must be conserved.) Traveling across a resistor with the current is a voltage drop, against the current is a voltage gain. Traveling across a battery from negative to positive is a voltage gain and from positive to negative is a voltage drop.

> **Sample Problem 3**

What is the current in the top path of the following section of a circuit?

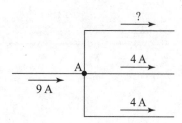

✓ **Solution**

Since the current coming into junction A on the left is 9 amps, a total of 9 amps must be leaving. Since the other two branches are carrying away 8 amps total, 1 amp is left for the missing pathway.

> **Sample Problem 4**

How much voltage drops across each resistor in this picture?

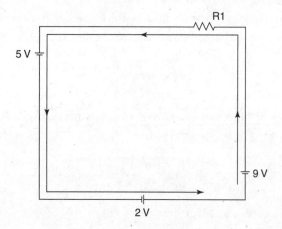

✓ **Solution**

If we trace a loop starting at the bottom right-hand corner and go counterclockwise, we encounter the following voltage changes, which must sum to zero:

$$+9\,V + R1 - 5\,V + 2\,V = 0$$

Solving for $R1$ gives us -6 V. This means current is going from right to left in this resistor as we traced through this resistor from right to left to obtain the decrease in voltage result. (Recall that current goes from high voltage to low voltage.)

Series Circuits

A series circuit consists of two or more resistors sequentially placed within one circuit. More generally, elements on the same path within a more complicated circuit are "in series with each other." An example is seen in Figure 3.6.

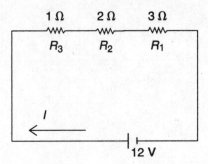

Figure 3.6 Series Circuit

We need to ask two questions. First, what is the effect of adding more resistors in series to the overall resistance in the circuit? Second, what effect does adding the resistors in series have on the current flowing in the circuit?

One way to think about this circuit is to imagine a series of doors, one after the other. As people exit through one door, they must wait to open another. The result is to drain energy out of the system and decrease the number of people per second leaving the room. In an electric circuit, adding more resistors in series decreases the current (with the same voltage) by increasing the resistance of the circuit.

Algebraically, all of these observations can be summarized as follows. In resistors R_1, R_2, and R_3, we have currents I_1, I_2, and I_3. Each of these three currents is equal to each other's current and to the circuit current I:

$$I = I_1 = I_2 = I_3$$

However, the voltage across each of these resistors is less than the source voltage V. If all three resistors were equal, each voltage would be equal to one-third of the total source voltage. In any case, we have

$$\Delta V = \Delta V_1 + \Delta V_2 + \Delta V_3$$

Using Ohm's law ($\Delta V = IR$), we can rewrite this expression as

$$IR = I_1 R_1 + I_2 R_2 + I_3 R_3$$

However, the three currents are equal, so they can be canceled out of the expression. This leaves us with

$$R = R_1 + R_2 + R_3$$

In other words, when resistors are added in series, the equivalent resistance of the circuit increases as the sum of all the resistances. This fact explains why the current decreases as more resistors are added. In our example, the equivalent resistance of the circuit is

$$1\,\Omega + 2\,\Omega + 3\,\Omega = 6\,\Omega$$

Since the source voltage is 12 V, Ohm's law provides for a circuit current of $12/6 = 2$ A. The voltage across each resistor can now be determined from Ohm's law since each gets the same current of 2 A. Thus,

$$V_1 = (1)(2) = 2\text{ V}; \quad V_2 = (2)(2) = 4\text{ V}; \quad V_3 = (3)(2) = 6\text{ V}$$

The voltages in a series circuit add up to the total, and not unexpectedly we have

$$2\,\text{V} + 4\,\text{V} + 6\,\text{V} = 12\,\text{V}$$

It should be noted that, when batteries are connected in series (positive to negative), the effective voltage increases additively as well.

Parallel Circuits

A parallel circuit consists of multiple pathways connected from one point to another, all experiencing the same potential difference. An example of a parallel circuit is seen in Figure 3.7.

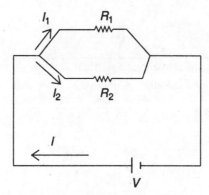

Figure 3.7 Parallel Circuit

In this circuit, a branch point is reached in which the current I is split into I_1 and I_2. Since each resistor is connected to a common potential point, experiments verify that the voltage across each resistor equals the source voltage ΔV. Thus, in this circuit it is the current that is shared; the voltage is the same. Another feature of the parallel circuit is the availability of alternative paths. If one part of the circuit is broken, current can flow through the other path. While each branch current is less than the total circuit current I, the effect of adding resistors in parallel is to increase the effective circuit current by decreasing the circuit resistance R.

To understand this effect further, imagine a set of doors placed next to each other along a wall in a room. Unlike the series connection (in which the doors are placed one after the other), the parallel-circuit analogy involves placing the doors next to each other. Even though each door will have fewer people per second going through it at any given time, the overall effect is to allow, in total, more people to exit from the room. This is analogous to reducing the circuit resistance and increasing the circuit current (at the same voltage).

Algebraically, we can express these observations as follows. We have two resistors, R_1 and R_2, with currents I_1 and I_2. Voltmeters placed across the two resistors would indicate voltages ΔV_1 and ΔV_2, which would be essentially equal to the source voltage ΔV; that is, in this example,

$$\Delta V = \Delta V_1 = \Delta V_2$$

Ammeters placed in the circuit would reveal that the circuit current I is equal to the sum of the branch currents I_1 and I_2; that is,

$$I = I_1 + I_2$$

Using Ohm's law, we see that

$$\frac{\Delta V}{R} = \frac{\Delta V_1}{R_1} + \frac{\Delta V_2}{R_2}$$

Since all voltages are equal, they can be canceled from the expression, leaving us with the following expression for resistors in parallel:

$$\frac{1}{R} = \frac{1}{R_1} + \frac{1}{R_2}$$

This expression indicates that the equivalent resistance is determined "reciprocally," which in effect reduces the equivalent resistance of the circuit as more resistors are added in parallel. If, for example, $R_1 = 10\ \Omega$ and $R_2 = 10\ \Omega$, the equivalent resistance is $R = 5\ \Omega$ (in parallel)! It should be noted that connecting batteries in parallel (positive to positive and negative to negative) has no effect on the overall voltage of the combination.

❯ Sample Problem 5

A 20 Ω resistor and a 5 Ω resistor are connected in parallel. If a 16 V battery is used, calculate the equivalent resistance of the circuit, the circuit current, and the amount of current flowing through each resistor.

✓ Solution

We know that the equivalent resistance is given by

$$\frac{1}{R_{eq}} = \frac{1}{R_1} + \frac{1}{R_2} = \frac{1}{20\ \Omega} + \frac{1}{5\ \Omega}$$

In this case, it is easily seen that $R_{eq} = 4\ \Omega$.

Thus, using Ohm's law,

$$I = \frac{\Delta V}{R_{eq}} = \frac{16\text{ V}}{4\ \Omega} = 4\text{ A}$$

To find the current in each branch, we recall that the voltage drop across each resistor is the same as the source voltage. In this case:

$$I_{20} = \frac{16\text{ V}}{20\ \Omega} = 0.8\text{ A}$$

$$I_5 = \frac{16\text{ V}}{5\ \Omega} = 3.2\text{ A}$$

Notice that the total current is equal to 4 A as expected.

Internal Resistance

Unless otherwise stated, on the AP Physics 2 exam, all wires, voltmeters, and batteries are assumed to be ideal (meaning they perform the same regardless of the conditions under which they are used). An ideal battery is one that has no internal resistance and thus delivers its specified voltage under all conditions. Real batteries, however, have some internal resistance, which means they will deliver less voltage at their terminals as more current is drawn. A battery that has become very hot or is near the end of its life will sometimes have significant internal resistance. Internal resistance is usually modeled as an ideal source of voltage (emf, ε, is the commonly used synonym for voltage in this model) in series with a small internal resistor (r), as in Figure 3.8. The voltage across the terminals of the battery ($\Delta V_{terminal}$, or the voltage difference from A to B in Figure 3.8) that is the useful voltage being delivered to the circuit is reduced by the voltage used by the internal resistance of the battery:

$$\Delta V_{terminal} = \varepsilon - Ir$$

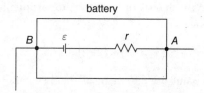

Figure 3.8 Internal Resistance

Combination Circuits

> **Sample Problem 6**

In Figure 3.9, a circuit that consists of resistors in series and parallel is presented. The key to reducing such a circuit is to decide whether it is, overall, a series or parallel circuit. In Figure 3.9, an 8 Ω resistor is placed in series with a parallel branch containing two 4 Ω resistors. The problem is to reduce the circuit to only one resistor and then to determine the circuit current, the voltage across the branch, and the current in each branch.

> Circuits with batteries of different potential in parallel will not appear on the AP Physics 2 exam.

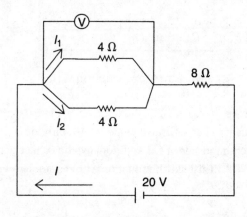

Figure 3.9

✓ **Solution**

To reduce the circuit and find the circuit resistance R, we must first reduce the parallel branch. Thus, if R_e is the equivalent resistance in the branch, then

$$\frac{1}{R_e} = \frac{1}{4} + \frac{1}{4}$$

This implies that $R_e = 2\ \Omega$ of resistance. The circuit can now be thought of as a series circuit between a 2 Ω resistor and an 8 Ω resistor. Thus,

$$R = 2\ \Omega + 8\ \Omega = 10\ \Omega$$

Since the circuit voltage is 20 V, Ohm's law states that the circuit current is

$$I = \frac{\Delta V}{R} = \frac{20}{10} = 2\ \text{A}$$

In a series circuit, the voltage drop across each resistor is shared proportionally since the same current flows through each resistor. Thus, for the branch,

$$\Delta V = IR_e = (2)(2) = 4 \text{ V}$$

To find the current in each part of the parallel branch, we recall that in a parallel circuit the voltage is the same across each resistor. Thus, since the two resistors are equal, each will get half of the circuit current; thus, each current is 1 A. Note how we used Kirchhoff's Rule.

Capacitors in Circuits

Capacitors in Series

We know from Chapter 2 that the capacitance of a parallel-plate capacitor is given by $C = Q/\Delta V$. Suppose we have a series of capacitors connected to a potential difference ΔV as shown in Figure 3.10.

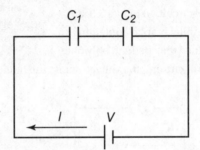

Figure 3.10 Capacitors in Series

In this circuit, the magnitude of the charge must be the same on all plates, so that portion must remain neutral. In order to remain neutral, any negative charges drawn to the right-hand side of C_1 will leave an equivalent amount of positive charge on the left-hand side of C_2. Additionally, the voltage across each capacitor is shared proportionally so that $\Delta V_1 + \Delta V_2 = \Delta V$ (just like the situation with resistors in series). Thus,

$$\frac{Q}{C_1} + \frac{Q}{C_2} = \frac{Q}{C}$$

and

$$\frac{1}{C_1} + \frac{1}{C_2} = \frac{1}{C} \quad \text{(for a series combination)}$$

Note that this is the opposite of the relationship for resistors in series. Be sure not to confuse them!

Capacitors in Parallel

A parallel combination of capacitors is shown in Figure 3.11.

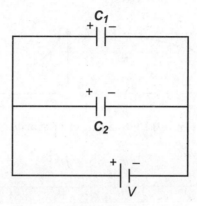

Figure 3.11 Capacitors in Parallel

In this circuit, the current branches off, and so each capacitor is charged to a different total charge, Q_1 and Q_2. The equivalent total charge is $Q = Q_1 + Q_2$, and the voltage across each capacitor is the same as the source emf voltage. Thus,

$$C\Delta V = C_1\Delta V + C_2\Delta V$$

and

$$C = C_1 + C_2 \quad \text{(for a parallel combination)}$$

Note that this is the opposite of the relationship for resistors in parallel.

Capacitors and Resistors in a Circuit

If there is both a capacitor and a resistor in a circuit, then there is a time-varying current as the capacitor "charges up" or "discharges." A typical RC circuit consists of a resistor and a capacitor connected in series together with a switch and a battery (see Figure 3.12).

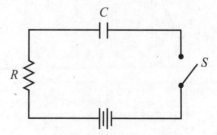

Figure 3.12 An RC Circuit

When the switch is closed, current flows through the resistor given by Ohm's law:

$$I = \frac{\Delta V_{\text{Battery}}}{R}$$

As the capacitor (with capacitance C) becomes charged, a potential difference (ΔV_C) appears across it because of the charge on the capacitor (Q_C), given by

$$\Delta V_C = \frac{Q_C}{C}$$

The current begins to decrease as the potential difference across the capacitor increases since there is less voltage available for the resistor. (Recall from Kirchhoff that $\Delta V_{Battery} = \Delta V_R + \Delta V_C$.) A graph of the increase in charge versus time for the capacitor is sketched in Figure 3.13.

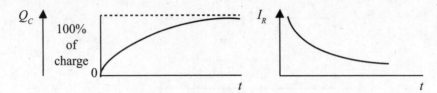

Figure 3.13 Charging of a Capacitor

> ## Sample Problem 7

What is the equivalent capacitance of a 5 F capacitor and a 15 F capacitor connected in parallel?

✓ ## Solution

We recall that in parallel, the total capacitance is equal to the sum of the individual capacitances:

$$C_T = C_1 + C_2 = 5\,F + 15\,F = 20\,F$$

On the AP Physics 2 exam, you will be required to solve steady-state RC circuits. So you must understand current and voltage everywhere once any charging or discharging of capacitors is complete. This simplifies matters considerably as a fully charged capacitor does not allow any current to flow through the pathway it is on as it can no longer accept any additional charges, while an uncharged capacitor initially behaves as if it were a simple wire. On the AP Physics 2 exam, however, you should know the qualitative aspect of the exponentially rising and falling nature of the current through and voltage across capacitors as they charge or discharge. The time scale (τ) over which this happens (specifically, it is the time it takes to gain or lose 63% of its charge during the charging or discharging process) can be characterized by the characteristic time constant of the RC circuit:

$$\tau = R_{eq} C_{eq}$$

TIP

As an exercise, use the definitions of the units *ohm* and *farad* to prove that, when multiplied, their product simplifies to the unit *second*!

> ## Sample Problem 8

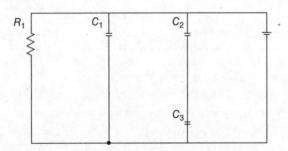

Given the values for the circuit components arranged as above, determine the steady-state solution for (a) charge on, (b) potential difference across, and (c) current through each component.

DC power supply = 9 V $R_1 = 4.5$ ohms $C_1 = C_2 = 18\,\mu F$ $C_3 = 9\,\mu F$

✓ Solution

Once the capacitors are fully charged, no current will be flowing through the central wires to the capacitors, allowing us to solve part (c) quickly:

$$I \text{ (for any of the capacitors)} = 0 \text{ A}$$

Since no current is flowing through any of the capacitors, the resistor and power supply are effectively in series.

$$I_{R_1} = I_{\text{power supply}} = \Delta V/R = 9/4.5 = 2 \text{ A}$$

Capacitor C_1 is in parallel with the power supply, so C_1 will receive the full 9 V (as does R_1). Capacitors C_2 and C_3 must split the 9 volts between them.

$$\Delta V_{R_1} = \Delta V_{C_1} = \Delta V_{C_2} + \Delta V_{C_3} = 9 \text{ V}$$

Resistors never accumulate charge, so $Q_{R_1} = 0$.

Knowing the voltage across C_1 allows us to solve for its charge quickly:

$$Q_{C_1} = C\Delta V = (18 \, \mu\text{F})(9 \text{ V}) = 162 \, \mu\text{C}$$

Recall that μ is simply the metric prefix *micro-*, which equals 10^{-6}. For C_2 and C_3, we must first determine their equivalent capacitance:

$$\frac{1}{C_{\text{equ}}} = \frac{1}{C_2} + \frac{1}{C_3} = \frac{1}{18} + \frac{1}{9}$$

$$C_{\text{equ}} = 6 \, \mu\text{F}$$

This equivalent capacitance would draw an amount of charge of:

$$Q = C\Delta V = (6 \, \mu\text{F})(9 \text{ V}) = 54 \, \mu\text{C}$$

Since this is the amount of charge drawn into the pathway containing both C_2 and C_3, the top plate of C_2 and the bottom plate of C_3 must receive exactly this much charge, giving us:

$$Q_{C_2} = Q_{C_3} = 54 \, \mu\text{C}$$

This result allows us to figure out how the 9 volts are distributed across C_2 and C_3:

$$\Delta V = Q/C$$

$$\Delta V_{C_2} = 54 \, \mu\text{C}/18 \, \mu\text{F} = 3 \text{ V}$$

$$\Delta V_{C_3} = 54 \, \mu\text{C}/9 \, \mu\text{F} = 6 \text{ V}$$

Electrical Energy Is Potential Energy

Many students of physics can get confused about the basic fact that the current going into a resistor is the same as the current coming out of the resistor. How can this be when the electrons have lost energy as the resistor heated up or gave off light? The conflict arises because many students are imagining the energy of the moving electrons as being kinetic energy. The energy in circuits, however, is electric potential energy! The electric potential energy is carried in the electric and magnetic fields associated with the moving charges. The electrons are spaced differently going into the resistor than they are when they are coming out. Even though the number of electrons passing per second, and hence their average speed (drift velocity), is the same before and after, the fields associated with their relative positions are completely different. Recall that all potential energies are energies of physical relationship between two or more interacting objects. Just as five fully extended rubber bands moving past you at 5 mph have more total energy than the same five slack rubber bands moving past you at the same speed, the slightly closer-spaced electrons before they enter the resistor have more energy than the slightly farther apart electrons leaving the resistor.

The purest example of this field energy for electricity is in electromagnetic radiation (or light). The light travels without any charges (or mass for that matter). Yet the light is demonstrably carrying energy in its fields as it goes from one place to another. Parallels can be made with other potential energies. The gravitational potential of a rock held above the ground is not actually stored within the rock itself but, rather, in the gravitational field between the rock and Earth. The elastic energy of a stretched spring does not belong to the block at the end of the spring but is in the stretched or compressed spring itself. The nuclear energy released or absorbed in nuclear reactions is not contained in the individual nucleons but in the fields exchanged between them in the form of the strong nuclear forces binding them together.

Almost all of the energy-releasing reactions we rely on in life come from a system moving from a state of negative potential energy (a weak bond) to a more negative potential energy state (a stronger bond). See Figure 3.14.

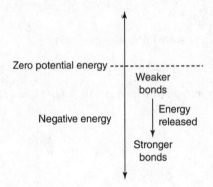

Figure 3.14 Binding Energies Are Negative

Examples:

GRAVITATIONAL ENERGY: A falling rock is gaining kinetic energy as it moves from being farther away from Earth to being closer to Earth (a strong gravitational bond).

CHEMICAL ENERGY: An exothermic reaction releases energy by moving from a weak chemical bond (small negative electric potential energy) to a stronger chemical bond (bigger negative electric potential energy).

NUCLEAR ENERGY: An energy-releasing fusion or fission reaction is one in which the daughter nuclei have stronger nuclear bonds (larger negative nuclear potential energies) than the original nuclei did.

SUMMARY

- Electric current is a measure of the flow of charge in units of amperes (1 A = 1 C/s).
- The conventional current is based on the direction of positive charge flow.
- Ohm's law states that at constant temperature the ratio of voltage and current is a constant in a conductor ($\Delta V/I = R$).
- Electrical resistance is based on the material used ("resistivity"), the length, and the cross-sectional area at constant temperature ($R = \rho L/A$).
- Ammeters measure current and are placed in series within the circuit.
- Voltmeters measure potential difference (voltage) and are placed in parallel (across segments of the circuit).
- A closed path and a source of potential difference are needed for a simple circuit.
- Resistors connected in series have an equivalent resistance equal to their numerical sum ($R_{eq} = R_1 + R_2$).
- Resistors connected in parallel have an equivalent resistance equal to their reciprocal sum $[1/R_{eq} = (1/R_1) + (1/R_2)]$.
- If capacitors are connected in series, then their equivalent capacitance is equal to their reciprocal sum (the opposite of resistors).
- If capacitors are connected in parallel, then their equivalent capacitance is equal to their numerical sum (the opposite of resistors).
- In the steady state of being fully charged, capacitors act as an open circuit stopping current from flowing down their pathway.
- Kirchhoff's Rules describe the flow of current in circuit branches and the changes in voltage around loops.

Problem-Solving Strategies for Electric Circuits

Several techniques for solving electric circuit problems have been discussed in this chapter. For resistances with one source of emf, we use the techniques of series and parallel circuits. Ohm's law plays a big role in measuring the potential drops across resistors and the determination of currents within the circuit.

When working with a combination circuit, try to reduce all sub-branches first as well as any series combinations. The goal is to be left with only one equivalent resistor and then use Ohm's law to identify the various missing quantities. Drawing a sketch of the circuit (if one is not already provided) is essential. Also, the direction of current is taken from the positive terminal of a battery. This is opposite to the way that electrons actually move!

Practice Exercises

1. How many electrons are moving through a current of 2 A for 2 s?

   (A) 3.2×10^{-19}
   (B) 4
   (C) 6.28×10^{18}
   (D) 2.5×10^{19}

2. How much electrical energy is generated by a 100 W light bulb turned on for 5 min?

   (A) 20 J
   (B) 500 J
   (C) 3,000 J
   (D) 30,000 J

3. What is the equivalent capacitance of the circuit shown below?

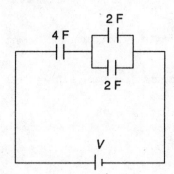

   (A) 2 F
   (B) 5 F
   (C) 8 F
   (D) 10 F

Questions 4 and 5 are based on the circuit below:

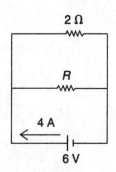

4. If the current in the battery is 4 A, then the power consumed by resistor R is

   (A) 6 W.
   (B) 7 W.
   (C) 18 W.
   (D) 24 W.

5. What is the value of resistor R in this circuit?

   (A) 3 Ω
   (B) 4 Ω
   (C) 5 Ω
   (D) 6 Ω

6. A 5 Ω and a 10 Ω resistor are connected in series with one source of emf of negligible internal resistance. If the energy produced in the 5 Ω resistor is X, then the energy produced in the 10 Ω resistor is

   (A) X.
   (B) $2X$.
   (C) $X/2$.
   (D) $X/4$.

7. Four equivalent light bulbs are arranged in a circuit as shown below. Which bulb will be the brightest?

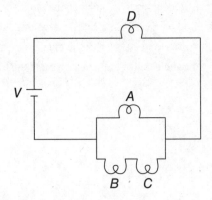

   (A) A
   (B) B
   (C) C
   (D) D

8. A 60 Ω resistor is made by winding platinum wire into a coil at 20°C. If the platinum wire has a diameter of 0.10 mm, what length of wire is needed?

9. Based on the circuit shown below:

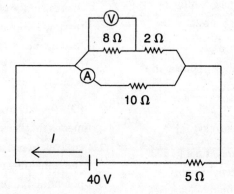

   (a) Determine the equivalent resistance of the circuit.
   (b) Determine the value of the circuit current *I*.
   (c) Determine the value of the reading of ammeter A.
   (d) Determine the value of the reading of voltmeter V.

10. Compare the brightness of two identical bulbs connected in series to their brightness when connected in parallel to the same battery. How does the power consumption of the battery compare?

11. You have five 2 Ω resistors. Explain how you can combine them in a circuit to have equivalent resistances of 1, 3, and 5 Ω.

12. A laboratory experiment is performed using light bulbs as resistors. After a while, the current readings begin to decline. What can account for this?

Answer Explanations

1. **(D)** By definition, 1 A of current measures 1 C of charge passing through a circuit each second. One coulomb is the charge equivalent of 6.28×10^{18} electrons. In this question, we have 2 A for 2 s. This is equivalent to 4 C of charge or 2.5×10^{19} electrons.

2. **(D)** Energy is equal to the product of power and time. The time must be in seconds; 5 min = 300 s. Thus

   $$\text{Energy} = (100)(300) = 30{,}000 \text{ J}$$

3. **(A)** We reduce the parallel branch first. In parallel, capacitors add up directly. Thus, $C_1 = 4$ F. Now, this capacitor would be in series with the other 4 F capacitor. In series, capacitors add up reciprocally; thus, the final capacitance is

   $$\frac{1}{C} = \frac{1}{4} + \frac{1}{4}$$

   This implies that $C = 2$ F.

4. **(A)** In a parallel circuit, all resistors have the same potential difference across them. The branch currents must add up to the source current (4 A, in this case). Thus, using Ohm's law, we can see that the current in the 2 Ω resistor is $I = 6/2 = 3$ A. Thus, the current in resistor R must be equal to 1 A. Since $P = \Delta VI$, the power in resistor R is

   $$P = (6)(1) = 6 \text{ W}$$

5. **(D)** Using Ohm's law, we have

   $$R = \frac{\Delta V}{I} = \frac{6}{1} = 6 \text{ Ω}$$

6. **(B)** In a series circuit, the two resistors have the same current but proportionally shared potential differences. Thus, since the 10 Ω resistor has twice the resistance of the 5 Ω resistor in series, it has twice the potential difference. Therefore, with the currents equal, the 10 Ω resistor will generate twice as much energy, that is, $2X$, when the circuit is on.

7. **(D)** The fact that all the resistors (bulbs) are equal means that the parallel branch will reduce to an equivalent resistance less than that of resistor D. The remaining series circuit will produce a greater potential difference across D than across that equivalent resistance. With the currents equal at that point, resistor D will generate more power than the equivalent resistance. Thus, with the current split for the actual parallel part, even less energy will be available for resistors A, B, and C. Bulb D will be the brightest.

8. The formula we want to use is

$$R = \frac{\rho L}{A}$$

The resistivity of platinum at 20°C is equal to

$$\rho = 10.4 \times 10^{-8} \, \Omega \cdot m$$

The resistance needed is $R = 60 \, \Omega$. We need the cross-sectional area. The diameter of the wire is 0.10 mm, so the radius of the wire must be equal to

$$r = \frac{\text{diameter}}{2} = 0.05 \text{ mm} = 0.00005 \text{ m}$$

The cross-sectional area is

$$A = \pi r^2 = \pi (0.00005 \text{ m})^2 = 7.85 \times 10^{-9} \text{ m}^2$$

The length of the wire is therefore

$$60 \, \Omega = \frac{(10.4 \times 10^{-8} \, \Omega \cdot m)L}{(7.85 \times 10^{-9} \text{ m}^2)}$$

$$L = 4.53 \text{ m}$$

9. (a) The first thing we need to do is reduce the series portion of the parallel branch. The equivalent resistance there is $8 \, \Omega + 2 \, \Omega = 10 \, \Omega$. Now, this resistance is in parallel with the other $10 \, \Omega$ resistor. The equivalent resistance is found to be $5 \, \Omega$ for the entire parallel branch. This equivalent resistance is in series with the other $5 \, \Omega$ resistor, making a total equivalent resistance of $10 \, \Omega$.

   (b) We can determine the circuit current using Ohm's law:

   $$I = \frac{\Delta V}{R} = \frac{40}{10} = 4 \text{ A}$$

   (c) To determine the reading of ammeter A, we must first understand that the potential drop across the $5 \, \Omega$ resistor on the bottom is equal to 20 V since

   $$\Delta V = IR = (4)(5) = 20$$

   Now, the equivalent resistance of the parallel branch is also $5 \, \Omega$, and therefore across the entire branch there is also a potential difference of 20 V (in a series circuit, the voltages must add up—in this case, to 40 V). In a parallel circuit, the potential difference is the same across each portion. Thus, the $10 \, \Omega$ resistor has 20 V across it. Using Ohm's law, we find that this implies a current reading of 2 A for ammeter A.

   (d) If, from part (c), ammeter A reads 2 A, the top branch must be getting 2 A of current as well (since the source current is 4 A). Thus, using Ohm's law, we can find the voltage across the $8 \, \Omega$ resistor:

   $$\Delta V = IR = (2)(8) = 16 \text{ V}$$

10. Bulb brightness is determined by the power consumed, $P = I\Delta V$, for each bulb.

    The two bulbs in series will have to split the voltage of the battery. Also, the two bulbs in series will receive half the current since they have doubled the resistance of their path. Combining these two effects, the power (brightness) of each bulb is one-fourth that of a single bulb attached to the same battery.

    The two bulbs in parallel will each have the full voltage of the battery and the full current of the battery. Therefore, each bulb burns just as brightly as a single bulb attached to the same battery.

    Since the bulbs in parallel are four times brighter, the battery will lose energy at four times the rate.

11. An equivalent resistance of 1 Ω can be achieved by connecting two 2 Ω resistors in parallel. An equivalent resistance of 3 Ω can be achieved by connecting two 2 Ω resistors in parallel and then connecting this system to one 2 Ω resistor in series. The addition of one more 2 Ω resistor in series to the combination above achieves an equivalent resistance of 5 Ω.

12. Light bulbs produce their luminous energy by heating up the filament inside them. Thus, after a while, this increase in temperature reduces the current flowing through them since the resistance has increased.

4

Magnetism and Electromagnetism

Learning Objectives

In this chapter, you will learn:

→ Magnetic fields and forces
→ Magnetic force on a moving charge
→ Magnetic fields due to currents in wires
→ Magnetic force between two wires
→ Induced motional emf in a wire
→ Magnetic flux and Faraday's law of induction

Magnetic Fields and Forces

When two statically charged point objects approach each other, each exerts a force on the other that varies inversely with the square of their separation distance. We can also state that one of the charges creates an electrostatic field around itself, and this field transmits the force through space.

When we have a wire carrying current, we no longer consider the effects of static fields. A simple demonstration reveals that, when we arrange two current-carrying wires parallel to each other so that the currents are in the same direction, the wires will be attracted to each other with a force that varies inversely with their separation distance and directly with the product of the currents in the two wires. If the currents are in opposite directions, the wires are repelled. Since this is not an electrostatic event, we must conclude that a different force is responsible.

The nature of this force can be further understood when we consider the fact that a statically charged object, such as an amber stone, cannot pick up small bits of metal. However, certain naturally occurring rocks called lodestones can attract metal objects. The lodestones are called **magnets** and can be used to **magnetize** metal objects (especially those made from iron or steel).

If a steel pin is stroked in one direction with a magnetic lodestone, the pin will become magnetized as well. If an iron rail is placed in the earth for many years, it will also become magnetized. If cooling steel is hammered while lying in a north-south line, it too will become magnetized (heating will eliminate the magnetic effect). If the magnetized pin is placed on a floating cork, the pin will align itself along a general north-south line. If the cork's orientation is shifted, the pin will oscillate around its original equilibrium direction and eventually settle down along this direction. Finally, if another magnet is brought near, the "compass" will realign itself toward the new magnet. This is illustrated in Figure 4.1.

Figure 4.1 Lining Up with Magnetic Fields

Each of these experiences suggests the presence of a magnetic force, and we can therefore consider the actions of this force through the description of a magnetic field ($\vec{B}$). If a magnet is made in the shape of a rectangular bar (see Figure 4.2), the orientation of a compass, moved around the magnet, demonstrates that the compass aligns itself tangentially to some imaginary field lines that were first discussed by English physicist Michael Faraday. In general, we label the north-seeking pole of the magnet as N and the south-seeking pole as S. We also observe that these two opposite poles behave in a similar way to opposite electric charges: like poles repel, while unlike poles attract.

A fundamental difference between these two effects is that magnet poles never appear in isolation. If a magnet is broken into two pieces, each new magnet has a pair of poles. This phenomenon continues even to the atomic level. There appear, at present, to be no magnetic monopoles naturally occurring in nature. (New atomic theories have postulated the existence of magnetic monopoles, but none has been discovered as of yet.)

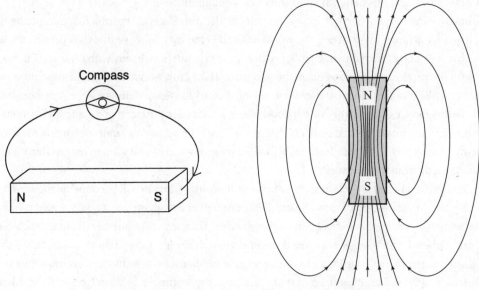

Figure 4.2 Magnetic Fields Around a Bar Magnet

Small pieces of iron (called "filings") can be used as miniature compasses to illustrate the characteristics of the magnetic field around magnets made of iron or similar alloys. The ability of a metal to be magnetized is called its permeability; substances such as iron, cobalt, and nickel have among the highest permeabilities. These substances are sometimes called **ferromagnets**.

REMEMBER

A compass needle will align itself tangent to the external magnetic field.

In Figure 4.3, several different magnetic field configurations are shown. By convention, the direction of the magnetic field is taken to be from the north pole of the magnet to the south pole. Note that magnetic field lines always form complete loops even if only the external fields are shown in pictures.

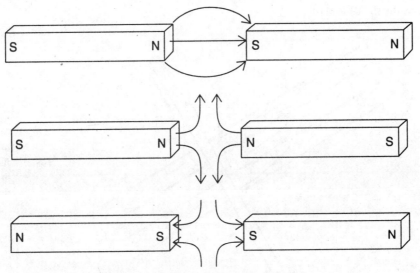

Figure 4.3 Magnetic Fields Between Magnets

To really understand the source of these magnetic fields and forces, we must examine the nature of the fields and forces themselves. The sources of all magnetic fields are moving charges. These magnetic fields, in turn, exert forces on only other moving charges. In the following sections, we will explore how magnetic forces are exerted on moving charges and then how the fields themselves are generated by moving source charges. Are the magnets discussed above somehow different? No, their magnetic properties arise from the microscopic movement of charges with the individual atoms within (specifically, both the intrinsic angular momentum of the electron itself and its angular momentum about the nucleus). Charges give rise to both electric and magnetic fields:

Charge	Produces
All charges	Electric fields
Moving charges	Magnetic fields
Accelerating charges	Changing electric and magnetic fields (electromagnetic radiation)

Magnetic Force on a Moving Charge

If a moving electric charge enters a magnetic field, it will experience a force that depends on its velocity, charge, and orientation with respect to the field. The force will also depend on the magnitude of the magnetic field strength (designated as B algebraically). The magnetic field is a vector quantity, just like the gravitational and electric field strengths.

Experiments with moving charges in a magnetic field reveal that the resultant force is a maximum when the velocity, v, is perpendicular to the magnetic field and zero when the velocity is parallel to the field. With this varying angle of orientation θ and a given charge Q, we find that the magnitude of the magnetic force is given by:

$$F = BQv \sin \theta$$

TIP

On the AP Physics 2 exam, only angles of 0°, 90°, and 180° will be used for calculations, although you should understand qualitatively what is happening at other angles.

For a current-carrying wire (I) of length ℓ:

$$F = BI\ell \sin\theta$$

Figure 4.4 illustrates the right-hand rule.

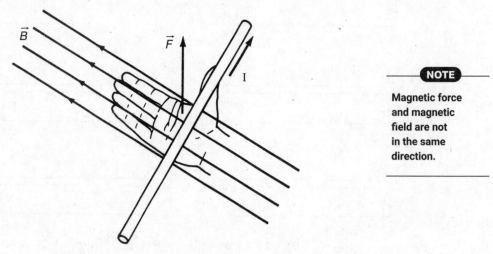

Figure 4.4 Magnetic Forces Are Perpendicular

> **NOTE**
>
> **Magnetic force and magnetic field are not in the same direction.**

❯ Sample Problem 1

A wire 1.2 m long and carrying a current of 60 A is lying at right angles to a magnetic field $B = 5 \times 10^{-4}$ T. Calculate the magnetic force on the wire.

✓ Solution

We use

$$F = BI\ell \sin(90°) = (5 \times 10^{-4}\text{ T})(60\text{ A})(1.2\text{ m})(1) = 0.036\text{ N}$$

Since the units of force must be newtons, the units for the magnetic field strength B must be newton · seconds per coulomb · meter (N · s/C · m). Recall that the newton · second is the unit for an impulse. We can think of the magnetic field as a measure of the impulse given to 1 coulomb of charge moving a distance of 1 meter in a given direction. Additionally, if we have a beam of charges, we effectively have an electric current. The units (seconds per coulomb) can be interpreted as the reciprocal of amperes, and so the strength of the magnetic field in a current-carrying wire is given in units of newtons per ampere · meter (N/A · m).

In the SI system, this combination is called a **tesla**; 1 tesla is equal to 1 newton per ampere · meter. Thus, an electric current carries a magnetic field. This is consistent with the earliest experiments by Hans Oersted, who first showed that an electric current can influence a compass. Oersted's discovery, in 1819, established the new science of **electromagnetism**.

The direction of the magnetic force is given by a "**right-hand rule**" (Figure 4.5): *Open your right hand so that your fingers point in the direction of the magnetic field and your thumb points in the direction of the velocity of the charges (or the current if you have a wire). Your open palm will show the direction of the force.*

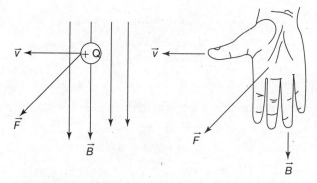

Figure 4.5 Right-Hand Rule for Magnetic Force

Figure 4.6 illustrates that, if the velocity and magnetic field are perpendicular to each other, the path of the charge in the field is a circle of radius R. The direction, provided by the right-hand rule, is the direction that a positive test charge would take. Thus, the direction for a negative charge would be opposite to what the right-hand rule predicted.

The circular path arises because the induced force is always at right angles to the deflected path when the velocity is initially perpendicular to the magnetic field. This path is similar to the circular path of a stone being swung in an overhead horizontal circle while attached to a string. Magnetic forces, like other centripetal forces, can do no work because of their perpendicularity.

The force vectors in Figure 4.5 are actually coming out of the page (straight toward the reader). It is hard to draw vectors unambiguously showing direction in this third direction. There is a convention to show arrows pointed straight out of the page and into the page:

⊙ means out of the page
⊗ means into the page

In Figure 4.6, the magnetic field is out of the page (pointed toward the reader), as indicated by the field of dots.

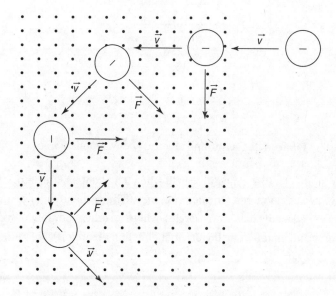

Figure 4.6 Magnetic Forces Can Be Centripetal

This relationship is given by:

$$F_c = ma_c = m\frac{v^2}{R} = BQv$$

where m is the mass of the charge.

This relationship can be used to determine the radius of the path:

$$R = m\frac{v}{QB}$$

We can use this expression to find the mass of the charge (as used in a mass spectrograph):

$$m = \frac{RQB}{v}$$

One way to determine the velocity of the charge, independent from the mass, is to pass the charge through both an electric and a magnetic field, as in a cathode ray tube. As shown in Figure 4.7, careful adjustment of both fields (keeping the velocity perpendicular to the magnetic field) can lead to a situation where the effect of one field balances out the effect from the other and the charges remain undeflected. In that case, we can write that the magnitudes of the electric and magnetic forces are equal ($F_E = F_B$):

$Eq = Bqv$, which implies that the velocity $v = E/B$ (independent of mass)!

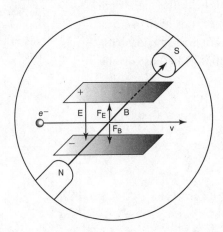

Figure 4.7 Combining Electric and Magnetic Forces

If the velocity is not perpendicular to the magnetic field when it first enters, the path will be a spiral because the velocity vector will have two components. One component will be perpendicular to the field (and create a magnetic force that will try to deflect the path into a circle), while the other component will be parallel to the field (producing no magnetic force and maintaining the original direction of motion). This situation is illustrated in Figure 4.8.

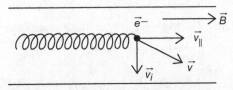

Figure 4.8 Spiraling in a Magnetic Field

> **Sample Problem 2**

An electron ($m = 9.1 \times 10^{-31}$ kg) enters a uniform magnetic field $B = 0.4$ T at right angles and with a velocity of 6×10^7 m/s.

(a)   Calculate the magnitude of the magnetic force on the electron.
(b)   Calculate the radius of the path followed.

✓ **Solution**

(a)   We use

$$F_m = Bqv$$

$$F_m = (0.4 \text{ T})(-1.6 \times 10^{-19} \text{ C})(6 \times 10^7 \text{ m/s}) = 3.8 \times 10^{-12} \text{ N}$$

(b)   We use

$$R = \frac{(mv)}{(Bq)} = \frac{(9.1 \times 10^{-31} \text{ kg})(6 \times 10^7 \text{ m/s})}{(0.4 \text{ T})(-1.6 \times 10^{-19} \text{ C})} = 8.5 \times 10^{-4} \text{ m}$$

> **Sample Problem 3**

What is the magnetic force (magnitude and direction) of a uniform length of wire of which only 50 cm is immersed in a uniform magnetic field, as shown below? The current is 0.75 A, while the magnetic field has a strength of 0.25 T. The wire and the magnetic field intersect at an angle of 105 degrees.

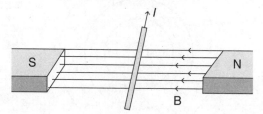

✓ **Solution**

For the magnitude, use the formula

$$F = BI\ell \sin \theta$$

$$F = (0.25)(0.75)(0.50) \sin(105°) = 0.091 \text{ N}$$

For the direction, use the right-hand rule to determine whether the force is upward or downward. (Note: Magnetic forces must always be perpendicular to both the field and the force. So up and down are the only two options in this case.)

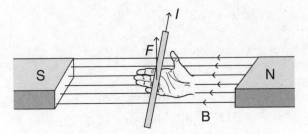

The magnetic force on the current-carrying wire is up.

Magnetic Fields Due to Currents in Wires

A Long, Straight Wire

If a compass is placed near a long, straight wire carrying current, the magnetic field produced by the current will cause the compass to align tangent to the field. Placing the compass at varying distances reveals that the strength of the field varies inversely with the distance from the wire and varies directly with the amount of current.

The shape of the magnetic field is a series of concentric circles (Figure 4.9) whose rotational direction is determined by a right-hand rule: *Grasp the wire with your right hand. If your thumb points in the direction of the current, your fingers will curl around the wire in the same direction as the magnetic field.*

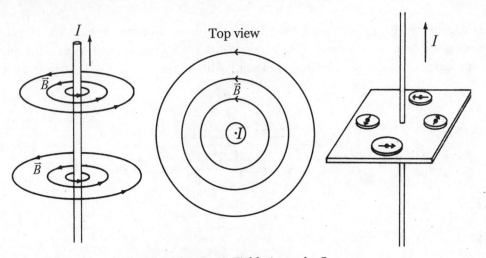

Figure 4.9 Magnetic Fields Around a Current

The magnitude of the magnetic field strength at some distance r from the wire is given by

$$B = \frac{\mu I}{2\pi r}$$

where μ is the permeability of the material around the wire in units of newtons per square ampere (Tm/A). In air, the value of μ is approximately equal to the permeability of a vacuum:

$$\mu_0 = 4\pi \times 10^{-7} \frac{\text{Tm}}{\text{A}}$$

This equation is therefore usually written in the form

$$B = \frac{\mu_0 I}{2\pi r}$$

Only μ_0 will be
used on the AP
Physics 2 exam.

A Loop of Wire

Imagine a straight wire with current that has been bent into the shape of a loop with radius r. Each of the circular magnetic fields now interacts in the center of the loop. In Figure 4.10, we see that the sum or superpositioning of all these contributing magnetic fields produces one concentrated field, pointing inward or outward in the middle.

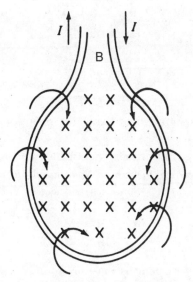

Figure 4.10 Right-Hand Rule for Currents

The general direction of the emergent magnetic field is given by a right-hand rule: *Point the thumb of your right hand in the direction of the current. Your fingers will curl in the direction of the magnetic field lines as they circle around the current.*

If the loop consists of N turns of wire, the magnetic field strength will be increased N times. The magnitude of the field strength, at the center of the loop (of radius r), is given by

$$B = \frac{\mu_0 N I}{2r}$$

The strength of the field can also be increased by changing the permeability of the core. If the wire is looped around iron, the field is stronger than if the wire were looped around cardboard. Also, if the loop's radius is smaller, the field is more concentrated at the center and thus stronger.

If the loop described in Figure 4.10 is stretched out to a length L, in the form of a spiral, we have what is called a **solenoid** (Figure 4.11). This is the basic form for an electromagnet, in which wire is wrapped around an iron nail and then connected to a battery. The strength of the field will increase with the number of turns, the permeability of the core, and the amount of current in the wire. The field inside the solenoid is uniform and given by:

$$B = \mu_0 \frac{N}{L} I$$

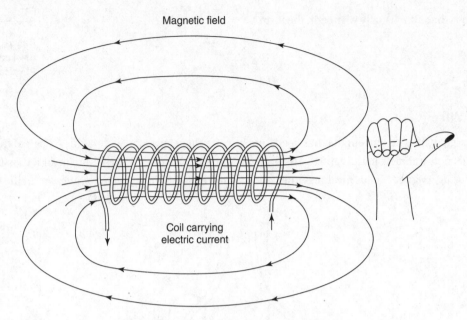

Figure 4.11 Right-Hand Rule for Solenoids

The direction of the magnetic field is given by another right-hand rule: *Grasp the solenoid with your right hand so that your fingers curl in the same direction as the current. Your thumb will point in the direction of the emergent magnetic field.*

> **Sample Problem 4**

What is the strength of the magnetic field at a point 0.05 m away from a straight wire carrying a current of 10 A?

✓ **Solution**

We use

$$B = \frac{\mu_0 I}{2\pi r} = \frac{(4\pi \times 10^{-7}\,\text{N/A}^2)(10\,\text{A})}{(2\pi)(0.05\,\text{m})} = 4 \times 10^{-5}\,\text{T}$$

Multiple Sources of Magnetism

What if there are several sources of magnetism in the same problem? The principle of superposition applies. Just like finding the net force, the net electric field, or the net gravitational field in a multiple-force problem, you must add (vector addition!) the fields from each source to find the net field at that location. This net magnetic field is the one used to determine any forces being applied to moving charges in the region. Recall that magnetic sources cannot apply magnetic forces to themselves!

Magnetic Force Between Two Wires

Imagine that you have two straight wires with currents going in the same direction (Figure 4.12). Each wire creates a circular magnetic field around itself so that, in the region between the wires, the net effect of each field is to weaken the other field relative to the other side. Although the resulting field around the two wires is found by adding the two magnetic fields produced by the two wires, each wire feels a force due to the other wire's magnetic field alone. Use the right-hand rule twice in a row to determine the force on each wire: thumb in direction of current, finger curl in direction of B-field. The first time is to determine the direction of the magnetic field in which the wire is immersed. The second time is to determine the force the wire feels in response to that magnetic field (see Figure 4.5). Note that the two wires form a Newton's third law pair, and their mutual magnetic forces are equal and opposite.

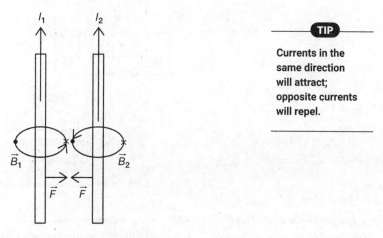

TIP

Currents in the same direction will attract; opposite currents will repel.

Figure 4.12 Magnetic Forces Between Currents

If the current in one wire is reversed, both forces will change directions, and the wires will repel each other.

The magnitude of the force between the wires, at a distance r, can be determined if we consider the fact that one wire sets up the external field that acts on the other. If we want the magnetic force on wire 2 due to wire 1, we can write

$$B_1 = \frac{\mu_0 I_1}{2\pi r}$$

Since $F = BIL$, we can write

$$F_{12} = \frac{\mu_0 I_1 I_2 L}{2\pi r}$$

where r is the distance between the wires.

Induced Motional emf in a Wire

We know from Oersted's experiments, mentioned previously, that a current in a wire generates a magnetic field. We also know that the field of a solenoid approximates that of an ordinary bar magnet. A third fact is that the magnetic property of a material (called its **permeability**) influences the strength of its magnetic field. Additionally, the interactions between fields lead to magnetic forces that can be used to operate an electric meter or motor.

A question can now be raised: Can a magnetic field be used to induce a current in a wire? The answer, investigated in the early nineteenth century by French scientist Andre Ampere, is yes. However, the procedure is not as simple as placing a wire in a magnetic field and having a current "magically" arise!

A simple experimental setup is shown in Figure 4.13. A horseshoe magnet is arranged with a conducting wire attached to a galvanometer. When the wire rests in the magnetic field, the galvanometer registers zero current. If the wire is then moved through the field in such a way that its motion "cuts across" the imaginary field lines, the galvanometer will register the presence of a small current. If, however, the wire is moved through the field in such a way that its motion remains parallel to the field, again no current is present. This experiment also reveals that a maximum current, for a given velocity, occurs when the wire moves perpendicularly through the field (which suggests a dependency on the sine of the orientation angle).

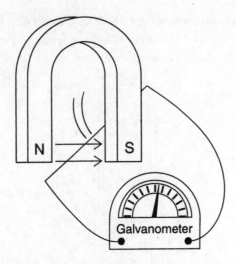

Figure 4.13 Induced Current from Motion

The phenomenon described above is known as **electromagnetic induction**. The source of the induced current is the establishment of a **motional emf** due to the change in something called the **magnetic flux**, which we will define shortly.

If the wire is moved up and down through the field in Figure 4.13, the galvanometer shows that the induced current alternates back and forth according to a right-hand rule. The fingers of the right hand point in the direction of the magnetic field; the thumb points in the direction of the velocity of the wire. Finally, the open palm indicates the direction of the induced current.

Now, since a parallel motion through the field implies no induced current, if the wire is moved at any other angle, only the perpendicular component contributes to the induction process. Recall that, if v is a perpendicular velocity component, $F = Bqv$ is the expression for the magnitude of the induced magnetic force on a charge q. Now, the wire in our situation contains charges that will experience a force F if we can get them to move through an external field. Physically moving the entire wire accomplishes this task. The direction of the induced force, determined by the right-hand rule, is along the length of the wire. If we let Q represent the total charge per second, we can write

$$F = BQv$$

The charges experience a force as long as they are in the magnetic field. This situation occurs for a length ℓ, and the work done by the force is given by $W = F\ell$. Thus

$$W = BQv\ell$$

The induced potential difference (emf) is a measure of the work done per unit charge (W/Q), which gives us

$$\text{emf} = B\ell v$$

This emf is sometimes called a "motional emf" since it is due to the motion of a wire through a field.

An interesting fact about electromagnetic induction is that the velocity in the previous expression is a relative velocity; that is, the induced emf can be produced whether the wire moves through the field or the field changes over the wire! Additionally, an emf can be induced even if there is no relative velocity as long as the magnetic field is changing.

Figure 4.14 is a simple illustration of this effect.

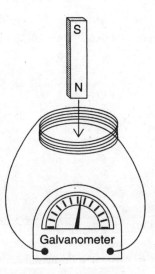

Figure 4.14 Induction

In Figure 4.14, a bar magnet is thrust into and out of a coil in which there are N turns of wire. The number of coils is related to the overall length ℓ in the previous equation. The galvanometer registers the alternating current in the coil as the magnet is inserted and withdrawn. Varying the velocity affects the amount of current as predicted. Moving the magnet around the outside of the coils generates only a weak current. The concept of relative velocity is illustrated if the magnet is held constant and the coil is moved up and down over the magnet. In summary, an emf (or Δv) can be induced by any of the following:

$$\Delta B, \Delta \theta, \text{ or } \Delta \ell$$

〉 Sample Problem 5

What is the induced emf in a wire 0.5 m long, moving at right angles to a 0.04 T magnetic field with a velocity of 5 m/s?

✓ Solution

We use

$$\text{emf} = B\ell v$$
$$\text{emf} = (0.04 \text{ T})(0.5 \text{ m})(5 \text{ m/s}) = 0.1 \text{ V}$$

Magnetic Flux and Faraday's Law of Induction

In Figure 4.15, there is a circular region of cross-sectional area A. An external magnetic field $\vec{B}$ passes through the region at an angle θ to the region. The perpendicular component of the field is given by

$$B_\perp = B\cos\theta$$

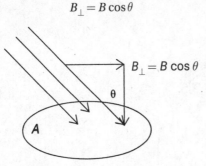

Figure 4.15 Magnetic Flux

> What is flux? Imagine the loop pictured in Figure 4.15 as the opening of a bucket and the field lines as streams of rain coming down. The flux would be the measure of how much water the bucket is actually capturing. Clearly, the amount of rain (field strength), the size of the opening (area), and the angle between them (determining the effective cross-sectional area) are the determining factors!

The magnetic flux, Φ, is defined as the product of the perpendicular component of the magnetic field and the cross-sectional area A:

$$\Phi = BA\cos\theta$$

The unit of magnetic flux is the **weber** (Wb); 1 weber equals 1 tesla per square meter.

On the basis of these ideas, the magnetic field strength is sometimes referred to as the **magnetic flux density** and can be expressed in units of webers per square meter (Wb/m^2).

English scientist Michael Faraday demonstrated that the induced motional emf (ε) was due to the rate of change of the magnetic flux. We now call this relationship **Faraday's law of electromagnetic induction** and express it in the following way:

$$\text{emf} = \varepsilon = -\left(\frac{\Delta\Phi}{\Delta t}\right)$$

The negative sign is used because another relationship, known as **Lenz's law**, states that an induced current will always flow in a direction such that its magnetic field opposes the magnetic field that induced it. Lenz's law is just another way of expressing the law of conservation of energy. Consider the experiment in Figure 4.14. When the magnet is inserted through the coil, an induced current flows. This current, in turn, produces a magnetic field that is directed either into or out of the coil, along the same axis as the bar magnet.

If the current had a direction such that the "pole" of the coil attracted the pole of the bar magnet, more energy would be derived from the effect than is possible in nature. Thus, according to Lenz's law, the current in the coil will have a direction so that the new magnetic field will oppose the magnetic field of the bar magnet. The more one tries to overcome this effect, the greater it becomes. The opposition of fields in this way prevents violation of the conservation of energy and can generate a large amount of heat in the process. If the flux is decreasing, the induced emf (voltage) will be such that its own magnetic field will add to the flux. If, on the other hand, the flux is increasing, the induced emf will be such that its own magnetic field will decrease the flux.

If there are N turns of wire in the coil, Faraday's law becomes:

$$\varepsilon = -N\left(\frac{\Delta\Phi}{\Delta t}\right)$$

❯ Sample Problem 6

A coil is made of 10 turns of wire and has a diameter of 5 cm. The coil is passing through the field in such a way that the axis of the coil is parallel to the field. The strength of the field is 0.5 T. What is the change in the magnetic flux? Also, if an average ε of 2 V is observed, for how long was the flux changing?

✓ Solution

The change in the magnetic flux is given by

$$\Delta \Phi = -BA = -B\pi r^2 = -(0.5)(3.14)(0.025)^2 = -9.8 \times 10^{-4}\,\text{Wb}$$

The time is therefore given by

$$\Delta t = N\left(\frac{\Delta \Phi}{\varepsilon}\right) = 10\left(\frac{9.8 \times 10^{-4}}{2}\right) = 4.9 \times 10^{-3}\,\text{s}$$

SUMMARY

- Magnets consist of both north and south poles.
- Isolated magnetic poles do not exist (no monopoles of magnets).
- The ability of a metal to become magnetized is called its permeability.
- Iron, cobalt, and nickel are metals with the highest permeabilities. Magnets made from these materials are called ferromagnets.
- Magnetic field lines go from north to south and loop back on themselves.
- An electric charge moving in an external magnetic field has a force induced on it. The direction of that force is determined by a right-hand rule.
- The path of a charged particle, traveling at right angles to an external magnetic field, is a circle.
- Metal wires carrying current generate magnetic fields. The magnetic field directions around wires can be found using a right-hand rule.
- Two wires carrying current can become attracted or repelled, depending on the directions of the currents.
- The forces induced on wires with currents can be used to make loops of wires spin in a motor or be controlled in an electric meter.
- A wire moving across an external magnetic field will have an emf induced in it.
- The induced ε will be a maximum if the wire cuts across the magnetic field at right angles to it.
- A wire moving parallel through an external magnetic field will not have an ε induced in it.
- Faraday's law states that the induced ε is equal to the rate of change of magnetic flux.
- Lenz's law states that an induced current will always flow in a direction such that its magnetic field opposes the change in magnetic flux that induced it.

Problem-Solving Strategies for Magnetism

The key to solving magnetic and electromagnetic field problems is remembering the right-hand rules. These heuristics have been shown to be useful in determining the directions of field interaction forces. Memorize each one and become comfortable with its use. Some rules are familiar in their use of fingers, thumb, and open palm. Make sure, however, that you know the net result. Drawing a sketch always helps.

Keeping track of units is also useful. The tesla is an SI unit and as such requires lengths to be in meters and velocities to be in meters per second (forces are in newtons). Most force interactions are angle dependent and have maximum values when two quantities (usually velocity and field or current and field) are perpendicular. Be sure to read each problem carefully and to remember that magnetic field $\vec{B}$ is a vector quantity!

Problem-Solving Strategies for Electromagnetic Induction

When solving electromagnetic problems, keep in mind that vector quantities are involved. The directions of these vectors are usually determined by the right-hand rules. Additionally, remember Lenz's law: **An induced current will always flow in a direction such that its magnetic field opposes the change in magnetic flux that induced it**, which plays a role in many applications. When solving problems in electromagnetic induction, also keep in mind that the induced emf is proportional to the change in the magnetic flux, not the magnetic field.

Practice Exercises

1. A charge moves in a circular orbit of radius R due to a uniform magnetic field. If the velocity of the charge is doubled, the orbital radius will become

   (A) $2R$.
   (B) R.
   (C) $R/2$.
   (D) $4R$.

2. Inside a solenoid, the magnetic field

   (A) is zero.
   (B) decreases along the axis.
   (C) increases along the axis.
   (D) is uniform.

3. An electron crosses a perpendicular magnetic field as shown below. The direction of the induced magnetic force is

   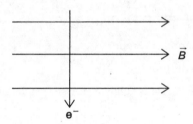

   (A) to the right.
   (B) to the left.
   (C) out of the page.
   (D) into the page.

4. Three centimeters from a long, straight wire, the magnetic field produced by the current is determined to be equal to 3×10^{-5} T. The current in the wire must be

   (A) 4.5 A.
   (B) 3 A.
   (C) 2.0 A.
   (D) 1.5 A.

5. Magnetic field lines determine

   (A) only the direction of the field.
   (B) the relative strength of the field.
   (C) both the relative strength and the direction of the field.
   (D) only the configuration of the field.

6. What is the direction of the magnetic field at point A above the wire carrying current?

   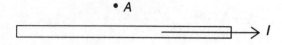

   (A) out of the page
   (B) into the page
   (C) up the page
   (D) down the page

7. A bar magnet is pushed through a flat coil of wire. The induced emf is greatest when

   (A) the north pole is pushed through first.
   (B) the magnet is pushed through quickly.
   (C) the magnet is pushed through slowly.
   (D) the south pole is pushed through first.

8. The magnetic flux through a wire loop is independent of

   (A) the shape of the loop.
   (B) the area of the loop.
   (C) the strength of the magnetic flux.
   (D) the orientation of the magnetic field and the loop.

9. A flat, 300-turn coil has a resistance of 3 Ω. The coil covers an area of 15 cm² in such a way that its axis is parallel to an external magnetic field. At what rate must the magnetic field change in order to induce a current of 0.75 A in the coil?

   (A) 0.0005 T/s
   (B) 0.0075 T/s
   (C) 2.5 T/s
   (D) 5 T/s

10. When a loop of wire is turned in a magnetic field, the direction of the induced emf changes every

    (A) one-quarter revolution.
    (B) two revolutions.
    (C) one revolution.
    (D) one-half revolution.

11. A wire of length 0.15 m is passed through a magnetic field with a strength of 0.2 T. What must be the velocity of the wire if an emf of 0.25 V is to be induced?

    (A) 8.3 m/s
    (B) 6.7 m/s
    (C) 0.12 m/s
    (D) 0.0075 m/s

12. An electron is accelerated by a potential difference of 12,000 V as shown in the following diagram. The electron enters a cathode ray tube that is 20 cm in length, and the external magnetic field causes the path to deflect along a vertical screen for a distance y.

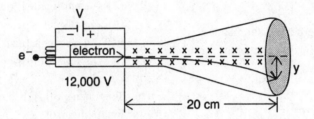

    (a) What is the kinetic energy of the electron as it enters the cathode ray tube?

    (b) What is the velocity of the electron as it enters the cathode ray tube?

    (c) If the external magnetic field has a strength of 4×10^{-5} T, what is the value of y? Assume that the field has a negligible effect on the horizontal velocity.

13. A straight conductor has a mass of 15 g and is 4 cm long. It is suspended from two parallel and identical springs as shown below. In this arrangement, the springs stretch a distance of 0.3 cm. The system is attached to a rigid source of potential difference equal to 15 V, and the overall resistance of the circuit is 5 Ω. When current flows through the conductor, an external magnetic field is turned on, and it is observed that the springs stretch an additional 0.1 cm. What is the strength of the magnetic field?

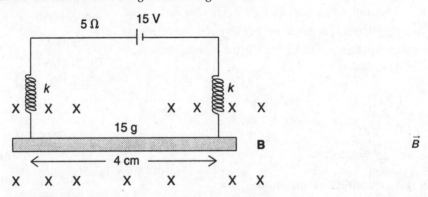

14. Explain why a bar magnet loses its magnetic strength if it is struck too many times.

15. A horizontal conducting bar is free to slide along a pair of vertical bars, as shown below. The conductor has length ℓ and mass M. As it slides vertically downward under the influence of gravity, it passes through an outward-directed, uniform magnetic field. The resistance of the entire circuit is R. Find an expression for the terminal velocity of the conductor (neglect any frictional effects due to sliding).

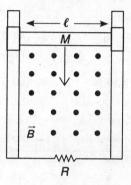

16. A small cylindrical magnet is dropped into a long copper tube. The magnet takes longer to emerge than the predicted free-fall time. Give an explanation for this effect.

Answer Explanations

1. **(A)** The formula is

$$R = \frac{mv}{qB}$$

   If the velocity is doubled, so is the radius.

2. **(D)** Inside a solenoid, the effect of all the coils is to produce a long, uniform magnetic field.

3. **(D)** Using the right-hand rule, we place the fingers of the right hand along the line of the magnetic field and point the thumb in the direction of the velocity. The palm points outward. However, since the particle is an electron, and the right-hand rule is designed for a positive charge, we must reverse the direction and say that the force is directed inward (into the page).

4. **(A)** The formula is

$$B = \frac{\mu_0 I}{2\pi r} = \frac{(2 \times 10^{-7})I}{r}$$

   Recalling that $r = 3$ cm $= 0.03$ m, we substitute all the given values to obtain $I = 4.5$ A.

5. **(C)** Magnetic field lines were introduced by Michael Faraday to determine both the direction of the field and its relative strength. (A stronger field is indicated by a greater line density.)

6. **(A)** As per the right-hand rule, point the thumb of your right hand to the right (in the direction of the current) and notice that your fingers will curve out of the page above the wire and into the page below the wire.

7. **(B)** The motional emf is equal to $B\ell v$, where v is the velocity of the bar magnet. Thus, the induced emf is greatest when the magnet is pushed through quickly.

8. **(A)** The magnetic flux is independent of the shape of the wire loop.

9. **(D)** The formula for the rate of change of magnetic flux is emf $= -N(\Delta\Phi/\Delta t)$. Using the given values and Ohm's law, we get for the magnitude of the flux change:

$$\left(\frac{\Delta\Phi}{\Delta t}\right) = \frac{(0.75)(3)}{300} = 0.0075 \text{ Wb/s}$$

   Our question concerns the rate of change of the magnetic field. In a situation in which the axis of the coil is parallel to the magnetic field, $\Phi = BA$, where $A = 0.0015$ m^2. Thus, $\Delta B/\Delta t = 5$ T/s.

10. **(D)** In an AC generator, the emf reverses direction every one-half revolution. The flux through the loop switches from getting bigger to getting smaller (Lenz's law).

11. **(A)** The motional emf is given by emf $= B\ell v$. Thus, using the given values, we find that $v = 8.3$ m/s.

12. (a) The kinetic energy of the electron is the product of its charge and potential difference. Thus,

$$KE = (1.6 \times 10^{-19})(12{,}000) = 1.92 \times 10^{-15} \, J$$

(b) The formula for the velocity, when the mass and kinetic energy are known, is

$$v = \sqrt{\frac{2KE}{m}} = \sqrt{\frac{(2)(1.92 \times 10^{-15})}{9.1 \times 10^{-31}}} = 6.5 \times 10^{7} \, m/s$$

(c) The magnetic force will cause a downward uniform acceleration, which can be found from Newton's second law, $F = ma$. The force will be due to the magnetic force, $F = qvB$. Thus, $a = qvB/m$. Using our known values, we get

$$a = 4.57 \times 10^{14} \, m/s^2$$

Now, for linear motion downward (recall projectile motion), $y = (1/2)at^2$. The time to drop y meters is the same time required to go 20 cm (0.20 m), assuming no change in horizontal velocity. Thus,

$$t = \frac{x}{v} = \frac{0.20}{6.5 \times 10^{7}} = 3 \times 10^{-9} \, s$$

This implies that, upon substitution, $y = 0.00205 \, m = 2.05 \, mm$.

13. Balancing the forces before the magnetic force is applied can help us determine the spring constant k:

$$2kx = mg$$

$$k = \frac{mg}{2x} = \frac{0.015(9.8)}{2(0.003)}$$

$$k = 24.5 \, N/m$$

Now balance the force of the additional stretch with the magnetic force applied to determine the strength of the magnetic field:

$$BIL = 2k(0.001)$$

$$B = \frac{2k(0.001)}{IL} = \frac{2(24.5)(0.001)}{3(0.04)}$$

$$B = 0.41 \, T$$

14. Striking a bar magnet or even heating it disrupts the magnetic domains that have aligned to magnetize it, causing the magnet to lose its strength.

15. Initially, the bar is accelerated downward by the force of gravity, given by $F = Mg$. The resistance R, in conjunction with the induced current I, produces an emf equal to IR and also to the product $B\ell v$. The magnetic force due to the current I is given by $BI\ell$. Combining all these ideas, we get that at terminal velocity:

$$Mg = I\ell B \quad \text{and} \quad I = \frac{B\ell v}{R}$$

Thus:

$$v_t = \frac{MgR}{\ell^2 B^2}$$

16. Copper is not highly magnetic. However, because of Lenz's law, the currents set up in the tube produce a magnetic field opposing the new field carried by the magnet. The result is to slow down the magnet's terminal velocity, as compared with the normal uniformly accelerated motion.

5

Waves, Sound, and Physical Optics

Learning Objectives

In this chapter, you will learn:

→ Pulses
→ Wave motion
→ Types of waves
→ Standing waves and resonance
→ Sound
→ The Doppler effect
→ Electromagnetic waves
→ Interference and diffraction of light
→ Polarization of light

Pulses

A pulse is a single vibratory disturbance in a medium. An example of a pulse is seen in Figure 5.1. If a string fixed at both ends and made taut, under a tension T (in newtons), is given a quick up and down snap, an upwardly pointing pulse is directed from the left to the right. The pulse appears to travel with a velocity v down the string. In actuality, the energy transferred to the string causes segments to vibrate up and down successively. This effect produces the illusion of a continuous pulse. Only energy is transferred by the pulse, and because the vibration is perpendicular to the apparent direction of motion, the pulse is said to be **transverse**. The **amplitude** of the pulse is the displacement of the disturbance above the level of the string.

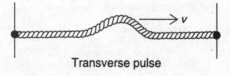

Transverse pulse

Figure 5.1

If the tension in the string is increased, the velocity of the pulse increases. If a heavier string is used (greater mass per unit length), the pulse appears to move slower. Careful measurements of these observations leads to the following equation for the velocity of the pulse:

$$v = \sqrt{\frac{T}{M/\ell}}$$

where T is the tension in newtons, M is the mass in kilograms, and ℓ is the length of the string in meters. In general, the velocity of pulses and waves is dependent on the **medium**. The medium is the material through which the waves and pulses are traveling.

When the pulse reaches one boundary, the energy transferred is directed upward against the wall, creating an upward force. Since the wall is rigid, the reaction to this action is a downward force. The string responds by being displaced downward and is reflected back. The result of this boundary interaction, illustrated in Figure 5.2, is observed as an inversion of the pulse upon reflection. Some energy is lost in the interaction, but most is reflected back with the pulse. This is known as "phase inversion."

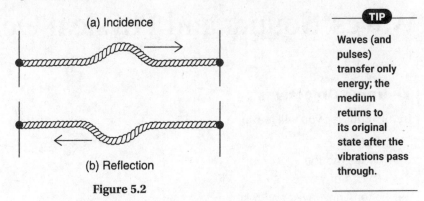

Figure 5.2

TIP

Waves (and pulses) transfer only energy; the medium returns to its original state after the vibrations pass through.

At a nonrigid boundary, say between two strings of different masses (but equal tensions), a different effect occurs. When the pulse reaches the nonrigid boundary, the upward displacement causes the second string to also be displaced upward. The magnitude of the second displacement depends on the mass density and tension in the second string. Thus, the transmitted pulse has an upward orientation and may travel with a larger or smaller velocity, depending, in this case, on the mass density.

The reflected pulse will be upward in orientation since no inversion takes place with a nonrigid boundary unless the difference between the two media (as measured by the second string's inertia) is large enough to behave as a semirigid boundary. As an example, consider a light string attached to a heavy rope, with the pulse traveling from the string to the rope. Figure 5.3 illustrates both cases.

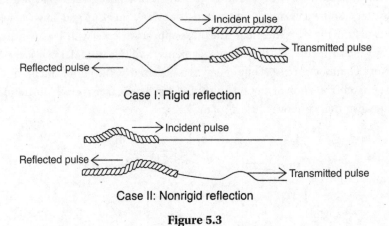

Figure 5.3

Since pulses transmit only energy, when two pulses interact, they obey a different set of physical laws than when two pieces of matter interact. Pulses are governed by the **principle of superposition**, which states that, when two pulses interact at the same point and at the same time, the interaction produces a single pulse whose amplitude displacement is equal to the sum of the displacements of the original two pulses. After the interaction, however, both pulses continue in their original directions of motion, unaffected by the interaction. Sometimes, this interaction is called **interference**.

When a pulse with an upward displacement interacts with another pulse with an upward displacement, the resulting pulse has a displacement larger than that of either original pulse (equal to the numerical sum of these pulses). This interaction is called **constructive interference**.

When a pulse with an upward displacement interacts with a pulse that has a downward displacement, the interaction is called **destructive interference**, and the resulting pulse has a smaller displacement than either original pulse (equal to the numerical difference of these pulses). Figure 5.4 illustrates these two types of interferences; *a* and *b* designate the two original amplitudes, and *p* is a point on the string.

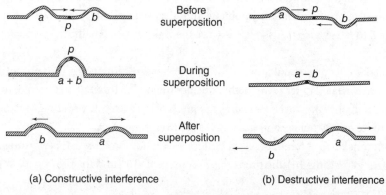

(a) Constructive interference   (b) Destructive interference

Figure 5.4

In the destructive interference case, it is possible, if the amplitudes are equal, that the interaction will momentarily cancel out both pulses. In any case, the orientation of the resulting pulse depends on the magnitude of each amplitude.

Wave Motion

If a continuous up and down vibration is given to the string in Figure 5.1, the resulting set of transverse pulses is called a **wave train** or just a **series of waves**. Whereas the pulse was just an upward- or downward-oriented, transverse disturbance, a wave consists of a complete up and down segment. Waves are thus periodic disturbances in a medium. The number of waves per second is called the **frequency**, designated by the letter *f* and expressed in units of reciprocal seconds (s^{-1}) or **hertz** (Hz). The time required to complete one wave cycle is called the **period**, designated by the capital letter *T* and expressed in seconds. The frequency and period of a wave are reciprocals of each other so that $f = 1/T$.

Like pulses, waves transmit only energy. The particles in the medium vibrate only up and down as the transverse waves approach and pass a given point. Since there is a continuous series of waves, many points in the medium will be moving up or down in unison. The **phase** of a wave is defined to be the relative position of a point on a wave with respect to another point on the same wave.

The distance between any two successive points in phase is a measure of the **wavelength**, designated by the Greek letter λ (lambda) and measured in meters. The peaks of the wave are called **crests**; the valleys, **troughs**. Thus, one wavelength can also be measured as the distance between any two successive crests or troughs. The **amplitude**, *A*, of the wave is the distance, measured in meters, of the maximum displacement, up or down, above the normal rest position (equilibrium level). Figure 5.5 illustrates a typical transverse wave.

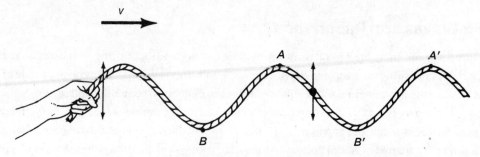

Figure 5.5 Transverse Wave

The sinusoidal nature of the wave is similar to the graph of displacement versus time for a mass on a spring undergoing simple harmonic motion. The maximum energy in a simple harmonic motion system is directly proportional to the square of the amplitude. This is indeed the case with a simple wave motion like the one shown in Figure 5.5. In this illustration we can identify points A and A' as all in phase. Points B and B' are also in phase, and the distance from B to B' can also be used to measure the wavelength. Finally, the velocity of the wave is given by the equation $v = f\lambda$.

Since each wave carries energy, it should not be surprising that the frequency is also proportional to the energy of a wave. The amplitude is related to what we might consider the wave's **intensity**. In sound, this property might be called **volume**. The frequency, however, measures the **pitch** of the wave. Waves of higher frequency transmit, at the same amplitude, more energy per second than waves of lower frequency.

There are many different kinds of frequencies that interact with human senses. It is worthwhile, therefore, to recall that the prefix *kilo-* represents 10^3, the prefix *mega-* represents 10^6, and the prefix *giga-* represents 10^9.

Types of Waves

In physics we usually deal with two types of waves, mechanical and electromagnetic. Waves that result from the vibration of a physical medium (a drum, a string, water, etc.) are called **mechanical waves**. Waves that result from electromagnetic interactions (light, X-rays, radio waves, etc.) are called **electromagnetic waves**. These electromagnetic waves are special since they do not require a physical medium to carry them. Also, all electromagnetic waves are transverse in nature.

Another way to set up periodic disturbances in the spring is to pull the spring back and forth along its longitudinal axis. The resulting disturbances consist of regions of compressions and expansions that appear to travel parallel to the disturbances themselves. These are called **longitudinal** waves (Figure 5.6) or **compressional** waves. Sound is an example of a longitudinal wave. The vibrations of an object in the air create pressure differences that alternately expand and contract the air and that, when impacted on our ears, create that phenomenon known as **sound**.

> **Reminder**
>
> Make sure you know the differences between **transverse** and **longitudinal waves**. You should also know the differences between **mechanical** and **electromagnetic waves**.

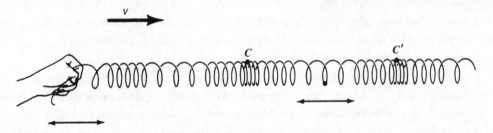

Figure 5.6 Longitudinal Wave

Standing Waves and Resonance

Consider Figure 5.1 again. There are several ways in which we can make the taut string vibrate transversely. One way is to make the entire string move up and down as a single unit. This is the easiest frequency mode of vibration possible and is called the **fundamental mode**. If we increase the frequency, something interesting happens. The waves reaching a boundary reflect off it inverted and match perfectly, in amplitude and frequency, the remaining incoming waves. In this situation, the apparent horizontal motion of the wave stops and we have a **standing wave**, which appears to be segmented by **nodal points**. These nodal points correspond to points where no appreciable displacement takes place.

Figure 5.7 shows a standing wave in various vibrational modes. Figure 5.7a illustrates the fundamental mode. In Figure 5.7b we see the second mode, in which one nodal point appears. This nodal point occurs at the one-half wavelength mark, so that, for a string of length ℓ, $\lambda = \ell$ (at this frequency). In the fundamental mode of Figure 5.7a, $\lambda = 2\ell$. In Figure 5.7c, there are two nodal points and $\lambda = 2\ell/3$.

This standing-wave pattern is a form of interference, and with sound the nodal points would be determined by a significant drop in intensity. With light, we would observe a darker region relative to the surrounding area. With a string, we would see the characteristic segmented pattern shown in Figure 5.7. Note that each segment is half a wavelength, bounded by two nodes with an **antinode** in the middle.

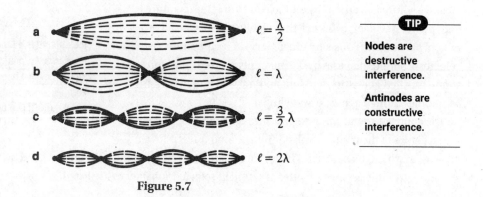

a $\ell = \dfrac{\lambda}{2}$

b $\ell = \lambda$

c $\ell = \dfrac{3}{2}\lambda$

d $\ell = 2\lambda$

Figure 5.7

> **TIP**
>
> **Nodes** are destructive interference.
>
> **Antinodes** are constructive interference.

Standing waves can be manipulated so that the maximum points continuously reinforce themselves. This buildup of wave energy due to the constructive interference of standing waves is called **resonance**. Resonance can also be induced by an external agent.

All objects have a natural vibrating frequency. When a glass or bell is struck, only one fundamental characteristic frequency of vibration is heard. (Actually, the sound is a complex mixture of vibrations.) If, however, two identical tuning forks, A and B, are held close together, then the phenomenon of resonance can be observed. If tuning fork A is struck, the waves emanating from it strike fork B. Since the frequency of these waves matches the natural vibrating frequency of fork B, that fork will begin to vibrate (see Figure 5.8).

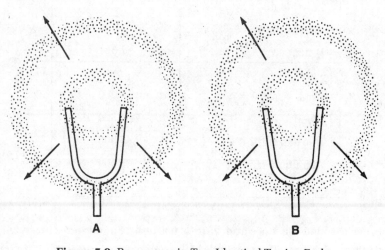

A   B

Figure 5.8 Resonance in Two Identical Tuning Forks

Resonance also explains how an opera singer can shatter a crystal wine glass. More important, it explains why soldiers are ordered to "break march" when crossing a bridge. The rhythmic marching can set up a resonance vibration and perhaps collapse the bridge. In November 1940, a resonance vibration caused by a light gale wind generated catastrophic torsional vibrations in the Tacoma Narrows Bridge in Washington State. The violent convulsions were photographed and are shown as a classic illustration of the principle of resonance. Nicknamed "Galloping Gertie," the bridge has become a testament to the need for engineers to be very careful when considering the possible effects of resonance.

Sound

Sound is a longitudinal mechanical wave. Air molecules are alternately compressed and spaced out (rarefied) as pressure differences move through the air. When lightning occurs, the expansion of air due to the very high temperature of the lightning bolt creates the violent sound known as thunder.

Since sound is "carried" by air molecules, it is temperature dependent. At 0 degree Celsius, the velocity of sound in air is 331 meters per second. For each 1-degree rise in temperature, the velocity of sound increases by 0.6 meter per second. The study of sound is called **acoustics**, and longitudinal waves in matter are sometimes termed **acoustic waves**.

> **Think About It**
>
> Since sound is a longitudinal wave, it cannot be polarized. Polarization can be used to test whether a wave has a transverse nature. For example, since light exhibits the property of polarization, it is a transverse wave.

The ability of sound waves to pass through matter depends on the structure of the matter involved. Thus, sound travels slower in a gas, in which the molecules have more random motion, than in a liquid or a solid, which has a more "rigid" structure. In Table 5.1 the velocities of sound in selected substances are listed.

Table 5.1 Speed of Sound in Selected Substances

Substance	Speed (m/s)
Gases (20°C)	
Carbon dioxide	267
Air	343
Helium	1,007
Liquids (25°C)	
Ethyl alcohol	1,207
Water, pure	1,498
Water, sea	1,531
Solids	
Lead	1,200
Wood	~4,300
Iron and steel	~5,000
Aluminum	5,100
Glass (Pyrex)	5,170

> **TIP**
>
> Wave *speed* is set by the medium.
>
> Wave *frequency* is set by the source.

Human hearing can detect sound in a range from 20 to 20,000 hertz. Sound waves exceeding a frequency of 20,000 hertz are called **ultrasonic** waves. The amplitude of the sound wave is the intensity of **loudness** of the wave, while the frequency relates to the **pitch**. When two sound waves interfere, the regions of constructive and destructive interference produce the phenomena known as **beats**. The number of beats per second is equal to the frequency difference.

Interference, for any kind of wave, is dependent on what is called the **path-length difference**. Consider two point sources of sound, A and B, and a receiver some distance away. If the distance from source A to the receiver is ℓ_A and the distance from source B to the receiver is ℓ_B, the receiver will be at a point of constructive interference if the path-length difference, $\ell_A - \ell_B$, is equal to a whole multiple of the wavelength, $n\lambda$. Destructive interference will occur if the path-length difference is equal to an odd multiple of the half-wavelengths, $(n + 1/2)\lambda$.

Another phenomenon associated with interference is **diffraction**. When a wave encounters a boundary, the wave appears to bend around the corners of the boundary. This effect, known as diffraction, occurs because at the corners the wave behaves like a point source and creates circular waves. These waves, because of their shape, reach behind the corners, and this continuous effect gives the illusion of wave bending. In Figure 5.9a, we observe a series of straight waves, made in a water tank, and the resulting diffraction at a corner. Figure 5.9b shows the diffraction of waves as they pass through a narrow opening. "Narrow" is relative to the wavelength of wave present. Waves do not show significant diffraction through openings much larger than their wavelength.

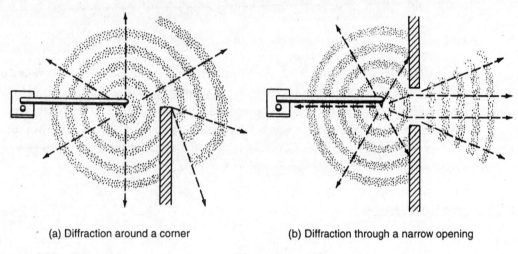

(a) Diffraction around a corner   (b) Diffraction through a narrow opening

Figure 5.9

When two sound waves have similar but slightly different frequencies (f_1 and f_2), they will constructively and destructively interfere such that humans will hear the change in amplitude over time. This is a commonly used phenomenon by musicians to tune instruments by ear. This interference phenomenon is known as the **beat frequency** and is simply the difference between the two sources:

$$|f_{\text{beat}}| = |f_1 - f_2|$$

The Doppler Effect

Almost everyone has had the experience of hearing a siren pass by. Even though the siren is emitting a sound at a single frequency, the changing position of the siren, relative to the hearer, produces an apparent change in the frequency. As the siren approaches, the pitch is increased; as the siren passes, the pitch is decreased. This phenomenon is called the **Doppler effect**.

Resonance, interference, diffraction, and the Doppler shift are phenomena associated with **ALL** waves!

In Figure 5.10 the source is moving; let's say toward point A. Then, for each period of time T between waves, the circular waves will not be concentric. In other words, the spacing between each two successive waves will be reduced by an amount equal to the distance traveled by the source in time T. Thus, an apparent increase in frequency is experienced at point A, while at point B there is an apparent decrease in frequency.

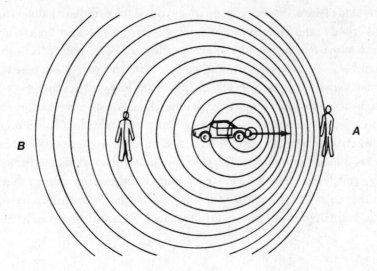

Figure 5.10

This Doppler shift, as it is known, is the result of the relative motion between the source of the wave and the receiver or observer. This can be accomplished by any combination of source and observer movement. As long as the two are approaching, the received frequency will be higher. When the two are getting farther apart, the received frequency will be lower. The amount of the shift in frequency is related to the relative velocity of the two compared with the velocity of the wave itself. Although there are precise equations associated with the Doppler effect, the AP Physics 2 exam only deals with this phenomenon qualitatively.

Electromagnetic Waves

In Chapter 4, we reviewed aspects of electromagnetic induction. In one example, we observed how the changing magnetic flux in a solenoid can induce an electric field. This **mutual induction** involves the transfer of energy through space by means of an oscillating magnetic field. Experiments in the late nineteenth century by German physicist Heinrich Hertz confirmed what Scottish physicist James Clerk Maxwell had asserted theoretically in 1864: that oscillating electromagnetic fields travel through space as transverse waves and that they travel with the same speed as does light. See Figure 5.11.

> **TIP**
>
> Electromagnetic waves do not need a medium through which to propagate. All electromagnetic waves travel with the velocity of light in a vacuum.

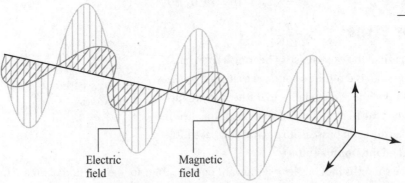

Electric field

Magnetic field

Figure 5.11 Electromagnetic Waves

Light is just one form of electromagnetic radiation that travels in the form of transverse waves. We know that these waves are transverse because they can be polarized (see the "Polarization of Light" section later in this chapter). The speed of light in a vacuum is approximately equal to 3×10^8 meters per second and is designated by the letter c.

Recall that every portion of a traveling wave can be modeled as a simple harmonic oscillator with its amplitude (A) and frequency (f) such that its displacement, x, has the motion of a sinusoidal function:

$$x = A \cos(2\pi f t)$$

Frequency, period (T), and angular frequency (ω) are all different ways of describing how quickly the oscillations happen. They are related and result in two additional ways of writing the previous function:

$$\omega = 2\pi f \quad \text{and} \quad f = 1/T$$

Experiments by Sir Isaac Newton in the seventeenth century showed that "white light," when passed through a prism, contains the colors red, orange, yellow, green, blue, and violet (abbreviated as ROYGBV). Each color of light is characterized by a different wavelength and frequency. The wavelengths range from about 3.5×10^{-7} meters for violet light to about 7.0×10^{-7} meters for red. For short wavelengths, the units **nanometers** (nm) are used; 1 nanometer is equal to 1×10^{-9} meters.

Unlike other electromagnetic waves, such as radio waves, X-rays, and infrared waves, light occupies a special place in the **electromagnetic spectrum** because we can "see" it. A sample electromagnetic spectrum is presented in Figure 5.12.

TIP

Make sure you
know the correct
order of the
electromagnetic
spectrum.
However, you will
not be required
to know the exact
ranges for the AP
Physics 2 exam.

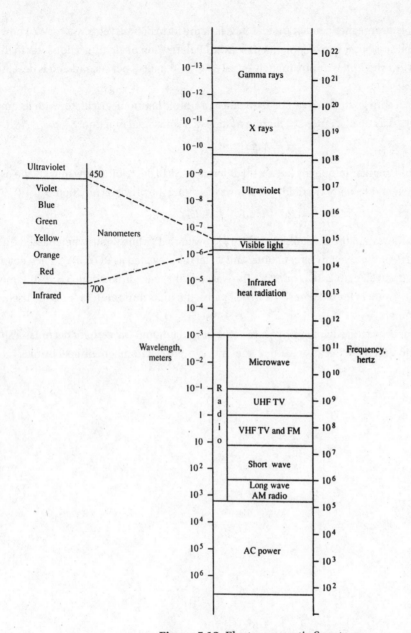

Figure 5.12 Electromagnetic Spectrum

Since electromagnetic waves are "waves," they all obey the relationship

$$c = f\lambda$$

where c is the speed of light, discussed earlier. With this in mind, gamma rays have frequencies in the range of 10^{25} hertz and wavelengths in the range of 10^{-13} meters. These properties make gamma rays very small but very energetic.

Interference and Diffraction of Light

The ability of light to diffract and exhibit an interference pattern is evidence of the wave nature of light. Light interference can be observed by using two or more narrow slits. Multiple-slit diffraction is achieved with a plastic "grating" onto which over 5,000 lines per centimeter are scratched.

Figure 5.13 shows that if white light is passed through the grating, a series of continuous spectra appears. Interestingly, this continuous spectrum is reversed from the way it appears in dispersion. The reason for this difference is that diffraction is wavelength dependent. Red, which has the longest wavelength, has to travel farther in order to experience constructive interference.

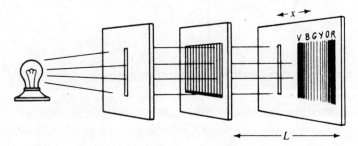

Figure 5.13 Multiple-Slit Diffraction

With monochromatic light, an alternating pattern of bright and dark regions appears. If a laser is used, the pattern appears as a series of dots representing regions of constructive and destructive interference (see Figure 5.14). The dots are evenly spaced throughout.

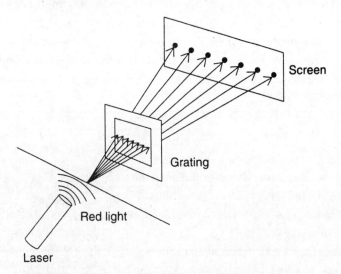

Figure 5.14 Monochromatic Diffraction Pattern

The origin of this pattern can be understood if we consider a water-tank analogy. Suppose two point sources are vibrating in phase in a water tank. Each source produces circular waves that overlap in the region in front of the sources. Where crests meet crests, there is constructive interference; where crests meet troughs, destructive interference. This situation is illustrated by the dotted arrows ($n = 0, 1, 2$) in Figure 5.15.

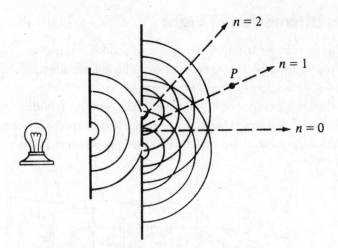

Figure 5.15 Double-Slit Interference

Notice in Figure 5.15 a central maximum ($n = 0$) built up by a line of intersecting constructive interference points. This line lies along a perpendicular bisector of the two slits, which serve as the two sources for the interference pattern. The symmetrical interference pattern consists of numbered "orders" that occur along evenly spaced angles above and below this central maximum.

In Young's double-slit experiment, the two explicit sources in the water tank are replaced by two narrow slits in front of a monochromatic ray of light. Each slit acts as a new point source of light that interferes with the waves from the other slit in much the same way as Figure 5.15 illustrates (this is known as **Huygens' principle**). The central maximum is called the "0-order maximum" because the **path-length difference** from the two sources (slits) is zero at this point.

Each successive constructive interference point ($n = 0$, 1, 2 in Figure 5.15) is caused by the path-length difference between the path from the higher slit (L_1) to that point of constructive interference and the path from the lower slit (L_2). Any time the difference in path-length is exactly 1 wavelength, the two waves arrive **in phase** and constructive interference occurs:

$$L_2 - L_1 = m\lambda$$

where m is an integer. This m value is precisely the number of the order in the diagram. Note that half-integer values give points of destructive interference (dark fringes) as the wave arrives at exactly half a wavelength or **out of phase**.

This path-length difference ($L_2 - L_1$) for the parallel slits interference pattern can be shown geometrically to be $d \sin\theta$ where d is the distance between slits and θ is the angle from the $n = 0$ ray in Figure 5.15 and the location x on the screen from Figure 5.13. The condition for interference fringes is now:

Don't forget that since light is a wave, it can also undergo Doppler shifts due to moving sources or receivers. However, since the speed of light is much larger than the speed of sound, Doppler shifts for light are not as common in our everyday lives. Speeds of cars and clouds, however, are frequently found via Doppler radar. (Radar is simply low-frequency light!)

$$d \sin\theta = m\lambda$$

If the angle is small enough (less than 10 degrees), then $\sin\theta$ may be approximated as the ratio of x/L, where x is the distance from the center to the mth fringe and L is the distance from the slits to the interference pattern:

$$d(x/L) = m\lambda$$

> ### Sample Problem 1

A ray of monochromatic light is incident on a pair of double slits separated by 8×10^{-5} m. On a screen 1.2 m away, a set of dark and bright lines appear separated by 0.009 m. What is the wavelength of the light used?

✓ Solution

We use

$$\lambda = \frac{dx}{L}$$

$$\lambda = \frac{(8 \times 10^{-5} \text{ m})(0.009 \text{ m})}{1.2 \text{ m}} = 6 \times 10^{-7} \text{ m}$$

Figure 5.16 shows another double-slit interference pattern.

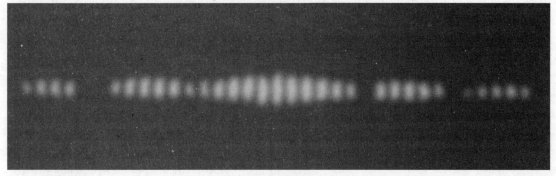

Figure 5.16

A careful examination of this double-slit interference pattern shows that, in addition to the expected regularly spaced interference fringes, there is a larger, superimposed pattern composed of a broad central maximum and several regularly spaced maxima and minima on either side. This is the interference of light with itself as it diffracts through an individual slit.

Single-slit interference patterns are governed by a similar formula (to that of double-slit patterns) for small angles (less than 10 degrees). However, in this formula, the a is the width of a single slit and y_{min} is the distance from the center to the mth minima on either side:

$$a\left(\frac{y_{min}}{L}\right) \approx m\lambda$$

Thin-Film Interference

Refer to Figure 5.17.

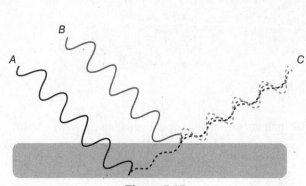

Figure 5.17

When light passes through and is reflected from the top layer (labeled *B*) and bottom layer (labeled *A*) of a thin film, the two reflected rays will interfere with each other (labeled *C*). In this context, a thin layer has a thickness comparable to the wavelength of light being used. Whether the light interferes constructively or destructively depends on two factors:

1. The usual path-length difference (twice the thickness of the layer)
2. The possible phase inversion of the wave at each reflection point

If twice the thickness of the thin film is equal to an integer number of wavelength, we expect constructive interference. However, if one of the two surfaces introduces a phase inversion, then this integer number of wavelength condition will give destructive interference. Similar to the phase inversion demonstrated in Figure 5.3, for light, the inversion happens when reflecting within a slower medium from a boundary with a faster medium. If both surfaces are phase inverting the light, then we go back to the constructive condition for integer numbers of wavelength.

Thin-film interference is responsible for many phenomena: the multicolored rainbow-like effect of a soap bubble in air, a similar pattern on a thin film of gasoline on a puddle, and minimizing reflected glare (or radar signals!) from surfaces. Note that there is usually an angle dependency for this pattern. On the AP Physics 2 exam, though, only light coming in perpendicular to the surfaces will be considered (so a path-length difference of simply twice the thickness is sufficient).

Polarization of Light

In a transverse wave, the vibrations must be perpendicular to the direction of apparent motion. However, if we cut a plane containing the directions of vibration, we find that there are many possible three-dimensional orientations in which the vibrations can still be considered perpendicular to the direction of travel. Thus, we can select one or more planes containing our chosen vibrational orientation. For example, a sideways vibration is just as "transverse" as an up-and-down or diagonal vibration. This selection process, known as **polarization** (see Figure 5.18), applies only to transverse waves.

In a longitudinal wave, there is only one way to make the vibrations parallel to the direction of travel. Special devices have been developed to test for polarization in mechanical and electromagnetic waves. Since all electromagnetic waves can be polarized, physicists conclude that they must be transverse waves. However, sound waves, being longitudinal, cannot be polarized.

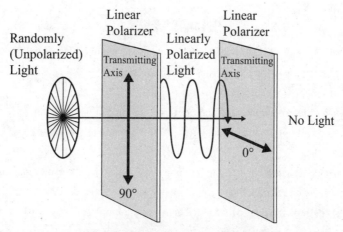

Figure 5.18 Polarization of Light

› Sample Problem 2

What are the advantages of wearing polarized sunglasses over wearing regular sunglasses? Support your answer in a paragraph format with supporting diagrams as needed.

✓ Solution

Both regular sunglasses and polarized sunglasses reduce the overall intensity of light entering your eyes, making it easier to see under bright conditions. However, the polarized sunglasses preferentially block the glare of reflected light from horizontal surfaces. The reason for this is that reflected light is partially (or, in some cases, completely) polarized horizontally compared with the unpolarized, unreflected light. By the use of a polarizing material on the lenses of the polarized sunglasses that blocks horizontally oriented light, not only will 50 percent of ambient light be eliminated but also the reflected light will be almost entirely eliminated as it is partially or completely polarized.

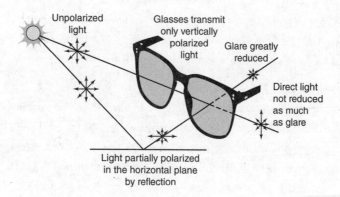

SUMMARY

- Waves and pulses can be transverse (vibration is perpendicular to the direction of travel, e.g., light) or longitudinal (vibration is parallel to the direction of travel, e.g., sound).
- Waves only transmit energy.
- The amplitude of a wave is its maximum displacement from its equilibrium position.
- The frequency of a wave is the number of cycles produced per second and is determined by the source of the wave.
- The period of a wave is the time it takes to complete one cycle and is the inverse of its frequency.
- The wavelength of a wave is the distance one complete cycle of the wave occupies.
- The speed of a wave is determined by the material it is passing through (the medium) and is equal to the product of its frequency times its wavelength.
- When waves meet, the principle of superposition applies (waves add together while overlapping and then pass through each other unchanged).
- Superposition of waves results in interference: constructive interference if the resulting amplitude is larger; destructive interference if the amplitude is smaller.
- A standing wave is an interference pattern created by an incident wave and the same wave reflected back onto itself. The nodes are points of destructive interference, while the antinodes are constructive interference. Node-to-node distance is half of a wavelength.
- Resonance is the transfer of energy from one body to another via waves of the object's natural vibrating frequency.
- The Doppler effect states that when a source and/or observer are moving relative to each other, there is an apparent change in frequency (and wavelength): higher frequency for relative motion toward each other; lower frequency for relative motion away from each other.
- Diffraction is the bending of waves around obstacles or emerging from openings in the same medium.
- Electromagnetic waves are produced by oscillating electromagnetic fields.
- Electromagnetic waves can travel through a vacuum and do not need a material medium for propagation.
- In a vacuum, all electromagnetic waves travel with the speed of light; $c = 3 \times 10^8$ m/s.
- Light rays travel in straight lines and produce shadows when incident on opaque objects.
- Luminous objects emit their own light. Opaque objects are seen by illumination; that is, they reflect light.
- The color of an opaque object is due to the selective reflection of certain colors.

Problem-Solving Strategies for Light

Remember that light is an electromagnetic wave. Therefore, it can travel through a vacuum. Reflection and refraction do not, by themselves, verify the wave nature of light. Diffraction and interference are evidence of the wave nature of light, and the ability of light to be polarized, using special Polaroid filters, is evidence that light is a transverse wave.

Keep in mind that, when light refracts, the frequency of the light is not affected. Since the velocity changes, so does the wavelength in the new medium. If the velocity in the medium is frequency dependent, the medium is said to be "dispersive." Light waves can be coherent if they are produced in a fashion that maintains constant phase-amplitude relationships. Lasers produce coherent light of one wavelength.

Practice Exercises

1. What is the frequency of a radio wave with a wavelength of 2.2 m?

   (A) $7.3 \times 10^{-9}\,\text{Hz}$
   (B) $1.36 \times 10^{8}\,\text{Hz}$
   (C) $2.2 \times 10^{8}\,\text{Hz}$
   (D) $3 \times 10^{8}\,\text{Hz}$

2. If the intensity of a monochromatic ray of light is increased while the ray is incident on a pair of narrow slits, the spacing between maxima in the diffraction pattern will be

   (A) increased.
   (B) decreased.
   (C) the same.
   (D) increased or decreased, depending on the frequency.

3. Widely recognized as a dangerous practice that should not be attempted, inhaling helium can temporarily change one's voice. If the wavelength of the standing wave determines the pitch heard, why does inhaling helium into one's larynx change the pitch of one's voice? Explain your answer.

4. (a) Light of wavelength 700 nm is directed onto a diffraction grating with 5,000 lines/cm. What are the angular deviations of the first- and second-order maxima from the central maxima?

   (b) Explain why X-rays, rather than visible light, are used to study crystal structure.

Answer Explanations

1. **(B)** The velocity of light in air is given by the formula $c = f\lambda$. The wavelength is 2.2 m, and the velocity of light in air is 3×10^8 m/s. Substituting known values gives us

$$f = 1.36 \times 10^8\,\text{Hz}$$

2. **(C)** The position and separation of interference maxima are independent of the intensity of the light.

3. Even though the size of one's vocal cords has not changed, the standing wave is actually created in the air within the larynx. Helium is a lighter molecule than those in regular air and thus has a higher wave speed than regular air. Since the wavelength is fixed by the size of one's vocal cords, and yet the wave speed has been increased, the frequency must go up.

4. (a) The general formula for diffraction is

$$N\lambda = d\sin\theta$$

where θ is the angle of deviation from the center. For the first-order maximum, $N = 1$, and d is equal to the reciprocal of the number of lines per meter. Thus, we must convert 5,000 lines/cm to 500,000 lines/m, and note that $d = 1/500{,}000$ meters. Now,

$$\sin\theta = \frac{N\lambda}{d} = \frac{(1)(7 \times 10^{-7})}{(1/500{,}000)} = 0.35$$

and

$$\theta = 20.5°$$

For the second-order maximum, we have $N = 2$. Thus

$$\sin\theta = \frac{(2)(7 \times 10^{-7})}{(1/500{,}000)} = 0.7$$

and

$$\theta = 44.4°$$

(b) X-rays are used to study crystal structure because of their very small wavelengths. These wavelengths are comparable to the spacings between lattices in a crystal and thus make it possible for the X-rays to be diffracted from the different layers. Visible light has wavelengths that are much greater than these spacings and so are not affected by them. The use of X-rays to probe crystals was one of the first diagnostic applications of these rays in atomic physics at the beginning of the twentieth century.

6

Geometric Optics

Learning Objectives

In this chapter, you will learn:

→ Reflection
→ Refraction
→ Total internal reflection
→ Image formation in plane mirrors
→ Image formation in curved mirrors
→ Image formation in lenses

Reflection

Reflection is the ability of light to seemingly bounce off a surface. There are two kinds of reflections: **specular** and **diffuse**. Specular reflections are reflections off of a smooth surface. They preserve the relative orientation of incoming light rays and are the primary type of reflection studied in physics. If the surface is rough and uneven, the reflection is said to be diffuse. The rays of incoming light are scattered in different directions. Note that each reflected light ray in a diffuse reflection is obeying the law of reflection. However, the overall effect is the light diffuses because of the microscopic variation in surface orientation. This phenomenon does not reveal the wave nature of light, and the notion of wavelength or frequency rarely enters into a discussion of reflection. If light is incident on a flat mirror, the angle of incidence is measured with respect to a line perpendicular to the surface of the mirror and called the **normal**.

In Figure 6.1, the **law of reflection**, which states that the angle of incidence equals the angle of reflection, is illustrated. Note that the angles of incidence and reflection and the normal are all coplanar.

> **Remember**
>
> The law of reflection states that the angle of incidence is equal to the angle of reflection. Remember that the angles are measured relative to the normal line.

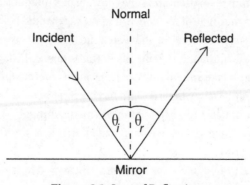

Figure 6.1 Law of Reflection

Reflection helps explain the colors of opaque objects. Ordinary light contains many different colors all blended together (so-called white light). A "blue" object looks blue because of the selective reflection of blue light due to the chemical dyes in the painted material. White paper assumes the color of the light incident on it because "white" reflects all colors. Black paper absorbs all colors (and of course nothing can be painted a "pure" color). If light of a single wavelength can be isolated (using a laser, for example), the light is said to be **monochromatic**. If all of the waves of light are moving in phase, the light is said to be **coherent**. Waves that match up peak to peak and trough to trough are said to be in phase (referring to the phase offset in sinusoidal functions). The fact that laser light is both monochromatic and coherent contributes to its strength and energy. (The word *laser* is an acronym for *l*ight *a*mplification by the *s*timulated *e*mission of *r*adiation.)

Refraction

Place a pencil in a glass of water and look at the pencil from the side. The apparent bending of the pencil is due to refraction (Figure 6.2a). As with all waves, wave speed depends on the medium. If light rays change speed as they pass from one medium to another medium, the light appears to bend toward or away from the normal. If there is no change in speed, as when light passes from benzene to Lucite, for example, there will be no refraction at any angle (Figure 6.2b). If the light is incident at an angle of zero degrees to the normal, there will again be no refraction, whether or not there is a change in speed (Figure 6.2c), since the entire wave front speeds up or slows down at the same time.

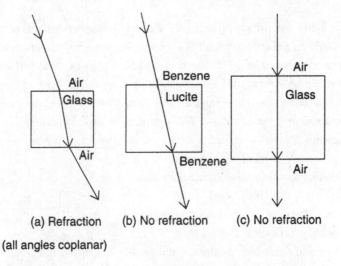

(a) Refraction   (b) No refraction   (c) No refraction

(all angles coplanar)

Figure 6.2 Refraction Requires Angle and Change in Speed

When light goes from one medium to another and slows down, at an oblique angle, the angle of refraction will be less than the angle of incidence, and we say that the light has been refracted **toward the normal**. Notice, in Figure 6.2a, that when the light reemerges into the air, it will be parallel to its original direction but slightly offset. This is due to the fact that the light is speeding up when it reenters the air. In that case, the angle of refraction will be larger than the angle of incidence, and we say that the light has been refracted **away from the normal**. Remember that, if the optical properties of the two media through which light passes are the same, no refraction occurs since there is no change in the speed of the light. In that case, the angle of incidence will be equal to the angle of refraction (no deviation).

> **TIP**
>
> **When light refracts, its frequency does not change.**

The extent to which a medium is a good refracting medium is measured by how much change there is in the speed of light passing through it. This "physical" characteristic is manifested by a "geometric" characteristic, namely, the angle of refraction. The relationship between these quantities is expressed by **Snell's law**. With the angle of incidence designated as θ_i and the angle of refraction as θ_r, and with v_1 the velocity of light in medium 1 (equal to c if medium 1 or 2 is air) and v_2 the velocity of light in medium 2, Snell's law states that

$$\frac{\sin \theta_i}{\sin \theta_r} = \frac{v_1}{v_2} = \frac{n_2}{n_1}$$

The absolute index of refraction, n, is defined by

$$n = c/v$$

where v is the wave speed in medium n and c is the speed of light in a vacuum (3.0×10^8 m/s). This gives us the usual formula for Snell's law:

$$n_1 \sin \theta_1 = n_2 \sin \theta_2$$

Note that when a change in medium dictates a change in speed, the light's frequency remains unchanged. The frequencies of all waves are dictated by the source of the wave. The wave speed and wavelength are determined by the medium (see Table 6.1). The light's wavelength is what changes (to a new λ_n) to accommodate the new velocity and to keep the wave equation true:

$$\lambda_n f = v$$

$$\lambda_n = \lambda/n$$

where λ is the wavelength in a vacuum.

Table 6.1 Absolute Indices of Refraction of Selected Media

Substance	Index of Refraction
Air (vacuum)	1.00
Water	1.33
Alcohol	1.36
Quartz	1.46
Lucite	1.50
Benzene	1.50
Glass	
Crown	1.52
Flint	1.61
Diamond	2.42

The particular colors of visible light all have specific frequencies. When these colors are used in a refraction experiment, they produce angles of refraction since they travel at different speeds in media other than air (or a vacuum). Substances that allow the frequencies of light to travel at different speeds are called **dispersive media**. This aspect of refraction explains why a prism allows one to see a colored, "continuous" spectrum and, additionally, why red light (at the low-frequency end) emerges on top (see Figure 6.3).

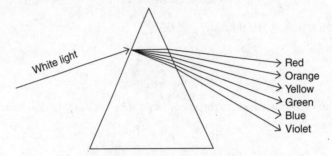

Figure 6.3 Prismatic Dispersion of Light

> **Sample Problem 1**

A ray of light is incident from the air onto the surface of a diamond ($n = 2.42$) at a $30°$ angle to the normal.

(a)   Calculate the angle of refraction in the diamond.

(b)   Calculate the speed of light in the diamond.

✓ **Solution**

(a)   We use Snell's law:

$$n_1 \sin \theta_1 = n_2 \sin \theta_2$$
$$(1.00) \sin(30°) = (2.42) \sin \theta_2$$
$$\theta_2 = 12°$$

(b)   We use

$$v = \frac{c}{n}$$

$$v = \frac{3 \times 10^8 \text{ m/s}}{2.42} = 1.2 \times 10^8 \text{ m/s}$$

Total Internal Reflection

When light is refracted from a medium with a relatively large index of refraction to one with a low index of refraction, the angle of refraction can be quite large.

Figure 6.4 illustrates a situation in which the angle of incidence is at some critical value θ_c such that the angle of refraction equals 90 degrees (ray D). This can occur only when the relative index of refraction is less than 1.00; i.e., the light is moving from a slower to a faster medium.

If the angle of incidence exceeds this critical value, the angle of refraction will exceed 90 degrees and the light will be 100% internally reflected, a phenomenon aptly called **total internal reflection** (ray E). The ability of diamonds to sparkle in sunlight is due to total internal reflection and a relatively small critical angle of incidence (due to a diamond's high index of refraction). Fiber optics communication, in which information is processed along hair-thin glass fibers, works because of total internal reflection.

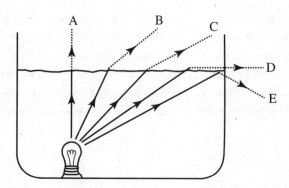

Figure 6.4 Critical Angle and Total Internal Reflection

The critical angle can be determined from the relationship:

$$\sin \theta_c = \frac{n_2}{n_1} \qquad (n_2 < n_1)$$

If $\theta > \theta_c$, total internal reflection occurs.

❯ Sample Problem 2

(a) Find the critical angle of incidence for a ray of light going from a diamond to air.

(b) Find the critical angle of incidence for a ray of light going from a diamond to water.

✓ Solution

(a) We use

$$\sin \theta_c = \frac{n_2}{n_1}$$

$$\sin \theta_c = \frac{1.00}{2.42} = 0.4132$$

$$\theta_c = 24°$$

(b) We again use

$$\sin \theta_c = \frac{n_2}{n_1}$$

$$\sin \theta_c = \frac{1.33}{2.42} = 0.5496$$

$$\theta_c = 33°$$

Image Formation in Plane Mirrors

When you look at yourself in a plane mirror, your image appears to be directly in front of you and on the other side of the mirror. Everything about your image is the same as you, the person, except for a left-right reversal. Since no light can be originating from the other side of the mirror, your image is termed **virtual**.

The formation of a virtual image in a plane mirror is illustrated in Figure 6.5. Using an imaginary point object, we construct two rays of light that emerge radially from the object because of ambient light from its environment. Each light ray is incident on the plane mirror at some arbitrary angle and is reflected off at the same angle (relative to the normal). Geometrically, we construct these lines using the law of reflection and a protractor. Since the rays diverge from the object, they continue to diverge after reflection. The human eye, however, perceives the rays as originating from a point on the other side of the mirror and in a direct line with the object.

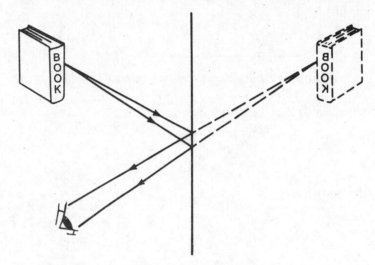

Figure 6.5 Plane Mirror

Image Formation in Curved Mirrors

If a concave or convex mirror is used, the law of reflection still holds, but the curved shapes affect the direction of the reflected rays. Figure 6.6a shows that a concave mirror converges parallel rays of light to a **focal point** that is described as **real** since the light rays really cross. In Figure 6.6b, we see that a convex mirror causes the parallel rays to diverge away from the mirror. If we extend the rays backward in our imagination, they appear to originate from a point on the other side of the mirror. This point is called the **virtual focal point** since it is not real. The human eye will always trace a ray of light back to its source in a line. This deception of the eye is responsible for images in some mirrors appearing to be on the "other side" of the mirror **(virtual images)**.

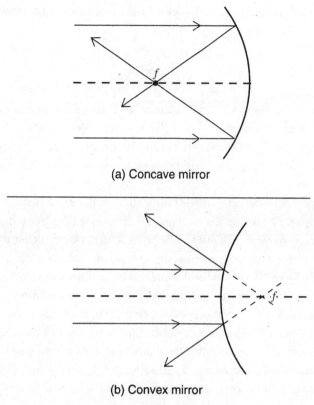

Figure 6.6 Basic Mirror Types

Concave Mirrors

In Figure 6.7, a typical concave mirror is illustrated. The axis is called the **principal axis**. If the curvature of the mirror is too large, a defect known as spherical aberration occurs and distorts the images seen. The real images formed by concave mirrors can be projected onto a screen. The focal length can be determined by using parallel rays of light and observing the point at which they converge.

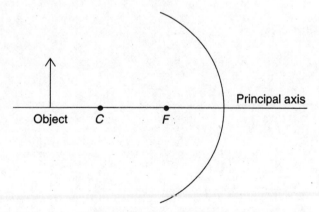

Figure 6.7 Concave Mirror

Another, more interesting, method is to aim the mirror out the window at distant objects. Again, the images of those objects can be focused onto a screen. Since the objects are very far away, they are considered to be **at infinity**. Normally, at infinity, an object sends parallel rays of light and appears as a point. Since we are viewing extended objects, we project a smaller image of them, and the distance from the screen to the mirror is the focal length F.

The point along the principal axis labeled C on Figure 6.7 marks the **center of curvature** (R = radius of curvature of the mirror) and is located at a distance $2F$.

$$F = \frac{R}{2}$$

To illustrate how to construct an image formed by a concave mirror, we use an arrow as an imaginary object. Its location along the principal axis is measured relative to points F and C. The orientation of the arrow is of course determined by the way it points. We could choose an infinite number of light rays that come off the object because of ambient light from its environment. For simplicity, however, we choose two rays that emerge from the top of the arrow.

In Figure 6.8, we present several different concave-mirror constructions. The first light ray is drawn parallel to the principal axis and then reflected through the focal point F. The second light ray is drawn through the focal point and then reflected parallel to the principal axis. The point of intersection (below the principal axis in cases I, II, and III) indicates that the image appears inverted at the location marked in the diagram.

From cases I, II, and III in Figure 6.8, we can see that the real images are always inverted. Additionally, as the object is moved closer to the mirror, the image gets larger and appears to move farther away from the mirror. Notice that, when the object is at point C, as in case II, the image is also at point C and is the same size. When the object is at the focal point, as in case IV, no image can be seen since the light is reflected parallel from all points on the mirror. If the object is moved even closer, as in case V, we get an enlarged virtual image that is erect. In case IV, a light ray has been drawn toward the center of the mirror; by the law of reflection, it will reflect below the principal axis at the same angle of incidence.

TIP

Make sure you know these cases for both curved mirrors and lenses.

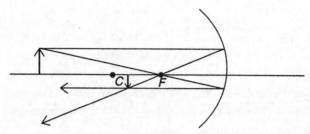

Case I—object beyond C; image is real, is between F and C, is smaller

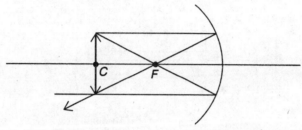

Case II—object at C; image is real, is at point C, is the same size

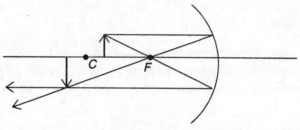

Case III—object between F and C; image is real, is beyond C, is larger

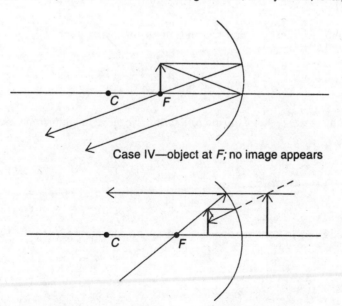

Case IV—object at F; no image appears

Case V—object between F and the mirror; image is virtual, is larger

Figure 6.8 Image Formation by Converging Mirror

Convex Mirrors

Convex mirrors are used in a variety of situations. In stores or elevators, for example, they have the ability to reveal images (although distorted) from around corners in aisles. An image in a convex mirror is always virtual and always smaller. This fact suggests that only one case construction, as opposed to five for concave mirrors, is necessary to understand image formation in these mirrors. In Figure 6.9, we show a sample construction and recall that convex mirrors diverge parallel rays. This divergence, however, is not at any arbitrary angle. The divergent ray is directed as though it originated from the virtual focal point.

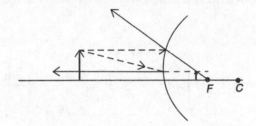

Image is virtual, upright, and smaller.

Figure 6.9 Image Formation by Diverging Mirror

Algebraic Considerations

We can study the images formed in curved mirrors by means of an algebraic relationship. We let f represent the focal length (positive in a concave mirror; negative in a convex mirror), S_o represent the object distance, and S_i represent the image distance, then, for a curved mirror. If the image is virtual, S_i will be negative.

$$\frac{1}{f} = \frac{1}{S_o} + \frac{1}{S_i}$$

Also, if we let h_o represent the object size and h_i represent the image size, then

$$m = \frac{h_i}{h_o} = \frac{-S_i}{S_o}$$

The ratio of image size to object size is called the **magnification**. Negative magnification means the image is inverted (upside down).

› Sample Problem 3

A 10 cm tall object is placed 20 cm in front of a concave mirror with a focal length of 8 cm. Where is the image located, and what is its size?

✓ Solution

Using the first formula presented, we can write:

$$\frac{1}{8} = \frac{1}{20} + \frac{1}{S_i}$$

Solving for the image distance, we obtain $S_i = 13.3$ cm. This is consistent with case I as illustrated in Figure 6.8.

The image size can be obtained from the expression:

$$h_i = h_o \left(\frac{S_i}{S_o} \right) = 10 \left(\frac{13.3}{20} \right) = 6.7 \text{ cm}$$

Image Formation in Lenses

As a further example of light refraction, consider the refraction due to a lens. If monochromatic light is used, Figure 6.10a demonstrates what happens when two prisms are arranged base to base and two parallel rays of light are incident on them. We can observe that the light rays converge to a real focal point some distance away. This situation simulates the effect of refraction by a double convex lens, as shown in Figure 6.10b.

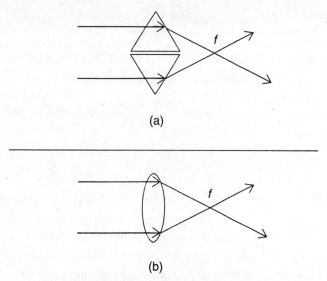

(a)

(b)

Figure 6.10 Converging Light Rays

The focal length is dependent on the frequency of light used; red light will produce a greater focal length than violet light in a convex lens.

If the prisms are placed vertex to vertex, as in Figure 6.11a, the parallel rays of light will be diverged away from an apparent virtual focal point. This simulates, as shown in Figure 6.11b, the effect of a double concave lens.

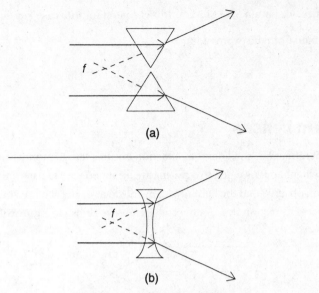

Figure 6.11 Diverging Light Rays

Note that in Figure 6.10, the focal point is said to be real because the light rays actually do cross there. In Figure 6.11, the focal point is virtual.

Converging Lenses

When discussing the formation of images in lenses, we usually invoke what is called the **thin lens approximation**; that is, we consider that the light begins to refract from the center of the lens. As a result, any curvature effects can be ignored. Since the top and the bottom of the lens are tapered like a prism, however, a defect known as chromatic aberration can sometimes occur. This defect causes the different colors of light to disperse in the lens and focus at different places because of their different frequencies and velocities in the lens material.

Figure 6.12 shows a series of cases in which a double convex lens is constructed as a straight line, using the thin lens approximation. The symmetry of the lens creates two real focal points, one on either side. Instead of considering the center of curvature (as in the curved mirror cases), we employ point $2F$ as an analogous location for reference. Our object is again an arrow drawn along the principal axis, which bisects the lens.

A light ray parallel to the principal axis will refract through the focal point. Therefore, a light ray originating from the focal point will refract parallel to the principal axis. We will use these two light rays to construct most of our images.

Notice that in case IV no image is produced, and we had to draw a different light ray passing through the optical center of the lens. In case V an enlarged virtual image is produced on the same side as the object. This is an example of a simple magnifying glass.

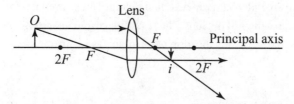

Case I–object beyond 2F; image is between F and 2F, is real, is smaller, is inverted

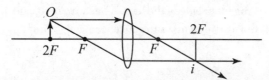

Case II–object at 2F; image is at 2F, is real, is same size but inverted

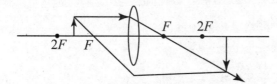

Case III–object between F and 2F; image is beyond 2F, is real, is larger, is inverted

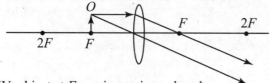

Case IV–object at F; no image is produced

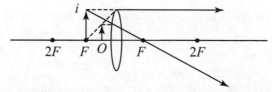

Case V–object between F and the lens; image is behind object, is virtual, is larger but upright

Figure 6.12 Image Formation by Converging Lens

Diverging Lenses

A concave lens will diverge parallel rays of light away from a virtual focal point. Figure 6.13 illustrates a typical ray construction for a diverging lens. Again, if we assume the thin lens approximation, we will draw the lens itself as a line for ease of construction. The context of the problem tells us that the drawing represents a concave, not a convex, lens.

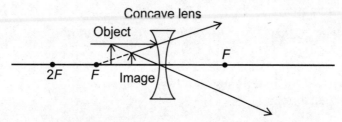

Figure 6.13 Image Formation by Diverging Lens

To construct the image, we have drawn a parallel light ray that then diverges away from the focal point. The second light ray is drawn through the optical center of the lens. The resulting virtual image will be erect and smaller and will be in front of the object. This will be true for any location of the object, so only one sample construction is necessary.

Algebraic Considerations

The relationship governing the image formation in a thin lens is the same as it was for a curved mirror. Positive image distances occur when the image forms on the other side of the lens. A negative image distance implies a virtual image. A positive focal length implies a convex lens, while a negative focal length implies a concave lens. Therefore, as before, we can write:

$$\frac{1}{f} = \frac{1}{S_o} + \frac{1}{S_i}$$

and

$$m = \frac{h_i}{h_o} = \frac{-S_i}{S_o}$$

If two converging lenses are used in combination, separated by some distance x, the combined magnification, m, of the system is given by:

$$m = m_1 m_2$$

❯ Sample Problem 4

A 3 cm tall object is placed 6 cm in front of a concave lens ($f = 3$ cm). Calculate the image distance and size.

✓ Solution

Algebraically rearrange the mirror lens equation above to solve for S_i:

$$S_i = \frac{(S_o f)}{(S_o - f)} \qquad \text{Recall that for a concave lens, } f < 0!$$

$$S_i = \frac{(6\text{ cm})(-3\text{ cm})}{6\text{ cm} - (-3\text{ cm})} = -2\text{ cm}$$

$$h_i = h_o\left(-\frac{S_i}{S_o}\right) = \frac{(3\text{ cm})(2\text{ cm})}{6\text{ cm}} = 1\text{ cm}$$

SUMMARY

	Sign of focal length	Object location	Image type/ orientation/sign	Image size
Converging systems (*convex lenses and concave mirrors*)	+	Beyond twice the focal distance	Real/Inverted/+	Smaller than object
		Between *f* and 2*f*	Real/Inverted /±	Bigger than object
		Inside the focal distance	Virtual/Upright/−	Bigger than object
Diverging systems (*concave lenses and convex mirrors*)	−	Any	Virtual/Upright/−	Smaller than object

- Reflection is the return of a wave in the opposite direction at the boundary of a new medium.
- Refraction is the pivot in the direction of a wave as it enters a new medium because of the change in wave speed.
- The law of reflection states that the angle of incidence is equal to the angle of reflection (as measured relative to a line normal to the surface).
- In optics, all angles are measured relative to the normal to a surface.
- Snell's law governs the refraction of light through transparent media.
- The ratio of the speed of light in air to the speed of light in a transparent medium is called the absolute index of refraction.
- The speed of light in a transparent medium can be obtained using Snell's law and is inversely proportional to the absolute index of refraction for the medium.
- Real images are formed when light rays converge. Real images can be projected onto a screen and are always inverted in appearance.
- Virtual images are formed as the brain imagines that reflected or refracted light rays converge back from their diverging paths. Virtual images cannot be projected onto a screen and are always erect. Virtual images are found by tracing the rays backward.
- Plane mirrors, convex mirrors, and concave lenses always produce virtual images. The virtual images produced by convex mirrors and concave lenses are always smaller than the object.
- Concave mirrors and convex lenses produce both real and virtual images. The real images may be larger or smaller than the object, whereas the virtual images are always larger.
- For a curved mirror, the focal length is equal to half the radius of curvature.
- If an object is placed at the focal point of a concave mirror or a convex lens, no image will be formed.

Problem-Solving Strategies for Geometric Optics

To solve ray-diagram problems, you must remember the types of light rays and understand how they reflect and refract using mirrors and lenses. Numerically, keep in mind that positive focal lengths imply concave mirrors and convex lenses.

Understand all of the cases illustrated in this chapter, and remember that real images are always inverted. Even if a given problem does not require a construction, draw a sketch. Additionally, remember that the focal length of a lens is dependent on the material of which the lens is made.

When doing refraction problems, remember that the light refracts toward the normal if it enters a medium of higher index of refraction and refracts away from the normal if it enters a medium of lower index of refraction. The absolute index of refraction of a substance can never be less than 1.00.

Practice Exercises

1. In the diagram below, a source of light (S) sends a ray toward the boundary between two media in which the relative index of refraction is less than 1. The angle of incidence is indicated by *i*. Which ray best represents the path of the refracted light?

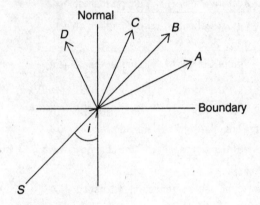

    (A) *A*
    (B) *B*
    (C) *C*
    (D) *D*

2. A coin is placed at the bottom of a clear trough filled with water ($n = 1.33$) as shown below. Which point best represents the approximate location of the coin as seen by someone looking into the water?

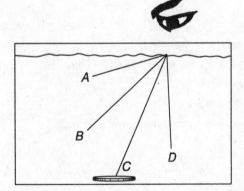

    (A) *A*
    (B) *B*
    (C) *C*
    (D) *D*

3. If, in question 2, the water is replaced by alcohol ($n = 1.36$), the coin will appear to be

    (A) higher.
    (B) lower.
    (C) the same.
    (D) higher or lower, depending on the depth of the alcohol.

4. An object is placed in front of a converging lens in such a way that the image produced is inverted and larger. If the lens were replaced by one with a larger index of refraction, the size of the image would

    (A) increase.
    (B) decrease.
    (C) increase or decrease, depending on the degree of change.
    (D) remain the same.

5. You wish to make an enlarged reproduction of a document using a copying machine. When you push the enlargement button, the lens inside the machine moves to a point

    (A) equal to *f*.
    (B) equal to 2*f*.
    (C) between *f* and 2*f*.
    (D) beyond 2*f*.

6. A negative image distance means that the image formed by a concave mirror will be

    (A) real.
    (B) erect.
    (C) inverted.
    (D) smaller.

7. Real images are always produced by

    (A) plane mirrors.
    (B) convex mirrors.
    (C) concave lenses.
    (D) convex lenses.

8. An object appears in front of a plane mirror as shown below:

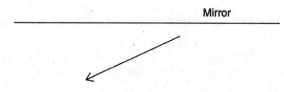

Which of the following diagrams represents the reflected image of this object?

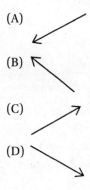

9. An object is located 15 cm in front of a converging lens. An image twice as large as the object appears on the other side of the lens. The image distance must be

(A) 15 cm.
(B) 30 cm.
(C) 45 cm.
(D) 60 cm.

10. An object is 25 cm in front of a converging lens with a 5 cm focal length. A second converging lens, with a focal length of 3 cm, is placed 10 cm behind the first one.

(a) Locate the image formed by the first lens.
(b) Locate the image formed by the second lens if the first image is used as an "object" for the second lens.
(c) What is the combined magnification of this combination of lenses?

11. Why do passenger-side mirrors on cars have a warning that states: "Objects are closer than they appear"?

12. Two lenses have identical sizes and shapes. One is made from quartz ($n = 1.46$), and the other is made from glass ($n = 1.5$). Which lens would make a better magnifying glass?

13. Why do lenses produce chromatic aberration whereas spherical mirrors do not?

14. Light is incident on a piece of flint glass ($n = 1.66$) from the air in such a way that the angle of refraction is exactly half the angle of incidence. What are the values of the angle of incidence and the angle of refraction?

15. A ray of light passing through air is incident on a piece of quartz ($n = 1.46$) at an angle of 25°, as shown below. The quartz is 1.5 cm thick. Calculate the deviation d of the ray as it emerges back into the air.

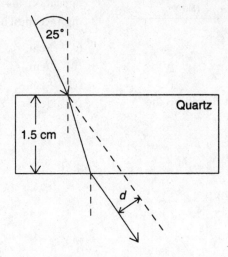

16. Immersion oil is a transparent liquid used in microscopy. It has an absolute index of refraction equal to 1.515. A glass rod attached to the cap of a bottle of immersion oil is practically invisible when viewed at certain angles (under normal lighting conditions). Explain how this might occur.

Answer Explanations

1. **(A)** Since the relative index of refraction is less than 1, the light ray will speed up as it crosses the boundary between the two media and will therefore bend away from the normal, approximately along path A.

2. **(B)** The human eye traces a ray of light back to its apparent source as a straight line. If we follow the line from the eye straight back, we reach point B.

3. **(A)** Since alcohol has a higher absolute index of refraction, the light will be bent farther away from the normal. Tracing that imaginary line straight back would imply that the image would appear closer to the surface (higher in the alcohol).

4. **(B)** The lens formula can be rewritten as

$$S_i = \frac{S_o f}{S_o - f}$$

If the object remains in the same position relative to the lens, then using a larger index of refraction will imply a smaller focal length. The image will move closer to the lens and, consequently, will be smaller.

5. **(C)** To produce an enlarged real image, the object must be located between f and $2f$.

6. **(B)** In a concave mirror, a negative image distance implies a virtual image, which is enlarged and erect.

7. **(D)** Only a convex lens or a concave mirror can produce a real image.

8. **(B)** The reflected image must point "away" from the mirror but on the other side, flipped over.

9. **(B)** The equation governing magnification in a converging lens is

$$M = \frac{S_i}{S_o}$$

Since $M = 2$ and $S_o = 15\,\text{cm}$, the image distance must be $S_i = 30\,\text{cm}$.

10. (a) For the first lens, we have $f = 5\,\text{cm}$ and $S_o = 25$; thus,

$$S_i = \frac{S_o f}{S_o - f} = \frac{(25)(5)}{25 - 5} = \frac{125}{20}$$

and the location of the image formed by the first lens is given by $S_i = 6.25\,\text{cm}$.

(b) Since $S = 10\,\text{cm}$ for the second lens (its distance from the first lens) and $S_i = 6.25$ for the first lens, the image q now serves as the new "object" at a distance $S'_o = 10 - 6.25 = 3.75$. The focal length of the second lens is $3\,\text{cm}$; therefore, the distance S'_i of the image formed by the second lens is

$$S'_i = \frac{(3.75)(3)}{3.75 - 3} = 15\ \text{cm}$$

(c) The combined magnification of the system is equal to the product of the separate magnifications. Since $m = S_i/S_o$, in general, we have

$$M_1 = \frac{6.25}{25} = 0.25$$

$$M_2 = \frac{15}{3.75} = 4$$

Therefore, the combined magnification of this combination of lenses is given by

$$M = M_1 M_2 = (0.25)(4) = 1$$

11. Passenger-side mirrors are convex in shape. This distorts images because of the divergence of the light, making objects appear to be smaller and more distant than they actually are.

12. Using the lens construction diagrams, we see that for case V the lens becomes a magnifying glass. Changing the focal length affects the size and location of the image. If the focal length is decreased, the image size and distance will increase. A shorter focal length is obtained by using a material with a higher index of refraction. Hence, glass would make a better magnifying glass.

13. Chromatic aberration is due to the dispersion of white light into the colors of the spectrum. Since a mirror does not disperse light on reflection, this problem does not occur. This observation was the motivation for Sir Isaac Newton to invent the reflecting telescope in 1671.

14. We want the angle of refraction to be equal to half the angle of incidence. This means that $\theta_i = 2\theta_r$. Now, since the light ray is initially in air ($n = 1.00$), Snell's law can be written:

$$\frac{\sin\theta_i}{\sin\theta_r} = n \text{ (glass)}$$

Substituting our requirement that $\theta_i = 2\theta_r$ gives

$$\frac{\sin 2\theta_r}{\sin\theta_r} = 1.66$$

Now, we recall the following trigonometric identity:

$$\sin 2\theta = 2\sin\theta\cos\theta$$

Thus,

$$\frac{2\sin r\cos r}{\sin r} = 1.66$$

and

$$2\cos r = 1.66$$

Therefore,

$$\cos r = 0.83$$

and

$$r = 34°$$

which means that

$$i = 68°$$

15. The diagram from the problem has been redrawn as shown below. From our knowledge of refraction, we know that angle θ must be equal to 25°. Angle r is given by Snell's law:

$$\frac{\sin 25^\circ}{\sin r} = 1.46$$

$$\sin r = \frac{\sin 25^\circ}{1.46} = 0.2894$$

$$r = 16.8^\circ$$

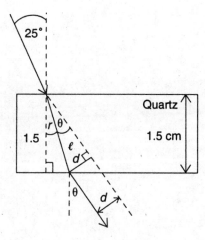

Now, angle θ is equal to the difference between the angle of incidence and the angle of refraction:

$$\theta = 25^\circ - 16.8^\circ = 8.2^\circ$$

Since the quartz is 1.5 cm in thickness, the length of the light ray, in the quartz, at the angle of refraction, can be determined from $\cos r = 1.5/\ell$. This implies that $\ell = 1.57$ cm. Now, since we know the length of the diagonal ℓ and the angle θ, the deviation d is just part of the right triangle in our diagram. Thus,

$$\sin \theta = \frac{d}{\ell}$$

and

$$d = \ell \sin \theta = (1.57) \sin 8.2^\circ = 0.224 \text{ cm}$$

16. The index of refraction for the glass is very nearly equal to that for the immersion oil. When the applicator rod is filled with liquid, the light passes through both media without refracting and makes the glass rod appear invisible.

7

Modern Physics

| Learning Objectives

In this chapter, you will learn:

- → Photoelectric effect
- → Blackbody radiation
- → Photon momentum
- → Matter waves
- → Spectral lines
- → Special relativity
- → Mass-energy equivalence
- → Atomic structure and Rutherford's model
- → The Bohr model
- → Quantum mechanics and the electron cloud model
- → Nuclear structure and stability
- → Binding energy
- → Radioactive decay
- → Fission
- → Fusion

Photoelectric Effect

Light has a very interesting effect on certain metals. Experimentally, it was observed that ultraviolet light could cause some metals to spark. Obviously, the light was causing electrons to be emitted from the metal. However, the basic mechanism was not understood. As we shall see, wave ideas about amplitude and frequency did not seem to make sense when applied to this experiment. Most troubling was the fact that raising the intensity of light (increasing the amplitude of the wave) could not cause electrons to be emitted if the frequency was too low. This phenomenon, called the **photoelectric effect**, defines a new view of light that has come to be known as the **quantum theory**. Developed by Albert Einstein, the quantum theory has been extended to particles as well as electromagnetic waves and forms the basis of modern physics.

The photoelectric effect can be studied in a more quantitative way. Figure 7.1 shows an evacuated tube that contains a photoemissive material. Many metals exhibit the photoelectric effect. On the other side of the tube is a collecting plate. The tube is connected to a microammeter, which can be used to measure the current generated by the emitted electrons. When exposed to light of suitable frequency, the ammeter will implicitly measure the number of electrons emitted per second. This number is sensitive only to the intensity of the electromagnetic wave applied. The minimum frequency that emits electrons is called the **threshold frequency** and is designated as f_0.

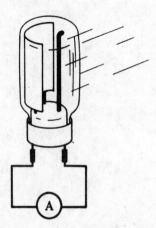

Figure 7.1 Vacuum Tube for Photoelectric Effect

To measure the energy of the emitted electrons, we insert in the circuit a source of potential difference (Figure 7.2) such that the negative electrons approach a negatively charged side. This will create a braking force; if we select the voltage properly, we can stop the current completely. The voltage that performs this feat is called the **stopping voltage** or **stopping potential** and is designated as V_0.

The energy of the electrons stopped by this voltage is equal to the product of the electrons' charge and the stopping voltage, eV_0. Since the stopping voltage is independent of the intensity of the light used (a violation of the wave theory of light), the stopping voltage measures the maximum kinetic energy of the emitted electrons.

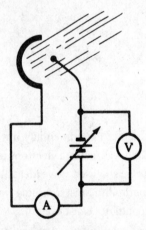

Figure 7.2 Photoelectric Effect Experimental Setup

As the frequency of light is slowly increased, there is a change in the maximum kinetic energy of the electrons. The fact that this energy is independent of the intensity of the light was used by Einstein to demonstrate that light consists of discrete or **quantized** packets of energy called **photons**. The energy of a given photon is directly proportional to its frequency.

To clarify this relationship, Figure 7.3 illustrates a typical graph of maximum kinetic energy versus frequency.

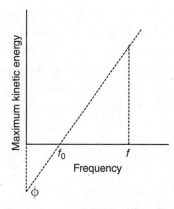

Figure 7.3 Results from a Photoelectric Effect Experiment

Starting with the threshold frequency, the graph is a straight line. For any frequency point, f, the slope of the line is given by

$$\text{Slope} = \frac{\text{KE}_{max}}{f - f_0} = h$$

This slope, referred to as **Planck's constant** and symbolized as h, has a value of

$$h = 6.63 \times 10^{-34}\ \text{J/Hz} = 6.63 \times 10^{-34}\ \text{J} \cdot \text{s}$$

Planck's constant originally appeared in the theoretical work of German physicist Max Planck. His mathematical discovery in 1900 of an elementary quantum of action helped explain the nature of radiation emitted from hot bodies. Planck's constant became a part of Einstein's quantum theory of light and Bohr's quantum theory of the hydrogen atom.

The above equation can be solved for the maximum kinetic energy:

$$\text{KE}_{max} = hf - \phi$$

In this expression, known as the **photoelectric equation**, hf represents the energy of the original photon, while ϕ, called the **work function**, is a property of the metal. The quantity ϕ is a measure of the minimum amount of energy needed to free an electron. The work function (ϕ) is sometimes designated as W_0 and can also be designated as hf_0. The threshold frequency (f_0) corresponds to the photon with the energy required to just barely free the electron.

Experiments with various metals yield some interesting results. Even though the different metals all have different threshold frequencies and, therefore, different work functions, they all obey the same photoelectric equation and have the same slope, h. A typical comparison graph of three metals, A, B, and C, is shown in Figure 7.4.

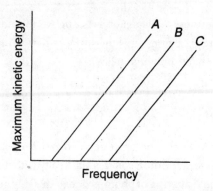

Figure 7.4 The Photoelectric Effect in Three Different Materials

Einstein was awarded the Nobel Prize for Physics in 1921 for his explanation of the photoelectric effect. His suggestion of the photoelectric effect, whereby light behaves as packets of quantized energy (photons), was in direct conflict with the prevailing wave theory of light, as demonstrated by interference and diffraction. Thus, physicists in the early twentieth century were presented with a **dual nature of light**. Generally speaking, the coexistence of particle and wave descriptions for fundamental particles is known as **wave-particle duality**.

> ### Sample Problem 1

Light with a frequency of 2×10^{15} Hz is incident on a piece of copper.

(a)   What is the energy of the light in joules and in electron volts?
(b)   If the work function for copper is 4.5 eV, what is the maximum kinetic energy, in electron volts, of the emitted electrons?

✓ Solution

(a)   The energy in joules is given by $E = hf$. Thus,

$$E = hf = (6.63 \times 10^{-34})(2 \times 10^{15}) = 1.326 \times 10^{-18} \text{ J}$$

Since $1 \text{ eV} = 1.6 \times 10^{-19}$ J, we have

$$E = 8.28 \text{ eV}$$

(b)   To find the maximum kinetic energy, we simply subtract the work function from the photon energy:

$$KE_{max} = 8.28 - 4.5 = 3.78 \text{ eV}$$

Blackbody Radiation

In resolving the photoelectric effect, Einstein was inspired by the earlier work of Max Planck, who had successfully explained the nature of light emitted by an object solely by virtue of its temperature. This emitted radiation is known as blackbody radiation (to differentiate it from the light that might sometimes be reflected from its surface). Planck's solution was to abandon classical concepts and quantize the energy in discrete chunks. In this way, he was able to successfully model the actual distribution of frequencies emitted by a blackbody. The peak emission intensity is found at a frequency proportional to the temperature (or inversely proportional to the wavelength) of an object (Wien's law):

$$\lambda_{max} = \frac{b}{T}$$

This is how we determine the temperature of distant objects, such as the surface of the Sun. The rate at which a blackbody emits electromagnetic energy is a function of the surface area (A) of the object and the temperature of the object raised to the fourth power (Stefan-Boltzmann law):

$$P = A\sigma T^4$$

Photon Momentum

In the early 1920s, American physicist Arthur Compton conducted experiments with X-rays and graphite. He was able to demonstrate that, when an X-ray photon collides elastically with an electron, the collision obeys the law of conservation of momentum. The scattering of photons due to this momentum interaction is known as the Compton effect (Figure 7.5). In the Compton effect, the scattered photon has a lower frequency than the incident photon.

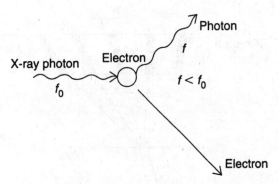

Figure 7.5 Compton Scattering

The momentum of the photon is given by the relationship $p = h/\lambda$. Thus X-rays, because of their very high frequency, possess a large photon momentum. Applying conservation of momentum with this expression for the momentum of a photon leads to the following expression of the change in wavelength of the photon from pre-collision to post-collision as a function of its angle of deflection:

$$\Delta\lambda = \frac{h}{m_e c}(1 - \cos\theta)$$

Even though we state that photons can have momentum, they do not possess what is called rest mass. If light can have momentum as well as wavelength, the question now raised is, can particles have wavelength as well as momentum?

Matter Waves

If a beam of X-rays is incident on a crystal of salt, the pattern that emerges is characteristic of the scattering of particles, not waves. If, however, a beam of electrons is incident on a narrow slit (Figure 7.6), the pattern that emerges is a typical wave diffraction pattern. Electron diffraction is one example of the dual nature of matter, which, under certain conditions, exhibits a wave nature. The circumstances are determined by experiment. If you perform an interference experiment with light, you will demonstrate its wave properties. If you do a photoelectric effect experiment, light will reveal its "particle-like" properties.

In the 1920s, French physicist Louis de Broglie used the dual nature of light to suggest that matter can have a "wave-like" nature, as indicated by its de Broglie wavelength. Since a photon's momentum is given by $p = h/\lambda$, the wavelength can be transposed to give $\lambda = h/p$.

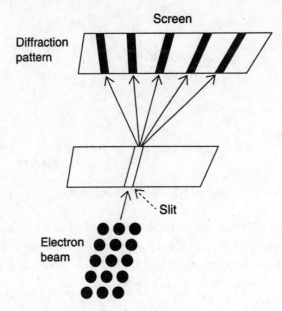

Figure 7.6 Interference from Electron Diffraction

Louis de Broglie suggested that, if the momentum of a particle, $p = mv$, has a magnitude suitable to overcome the small magnitude of Planck's constant, a significant wavelength can be observed. The criterion, since mass is in the denominator of the expression for wavelength, is that the mass be extremely small (on the atomic scale). Thus, while electrons, protons, and neutrons might have wave characteristics, a falling stone, because of its large mass, would not exhibit any wave-like effects. The equation for the de Broglie wavelength of a particle is given by

$$\lambda = \frac{h}{mv}$$

⟩ Sample Problem 2

Find the de Broglie wavelength for each of the following:

(a)   A 10 g stone moving with a velocity of 20 m/s.
(b)   An electron moving with a velocity of 1×10^7 m/s.

✓ Solution

(a)   Since $m = 10 \text{ g} = 0.01 \text{ kg}$, and $\lambda = h/mv$, we have

$$\lambda = \frac{6.63 \times 10^{-34}}{(0.01)(20)} = 3.315 \times 10^{-33} \text{ m}$$

(b)   In this part we have

$$\lambda = \frac{h}{mv} = \frac{6.63 \times 10^{-34}}{(9.1 \times 10^{-31})(1 \times 10^7)} = 7.3 \times 10^{-11} \text{ m}$$

This aspect of particle motion, on a small scale, leads to the so-called **Heisenberg uncertainty principle**. Developed by Werner Heisenberg, also in the 1920s, this principle states that because of the wave nature of small particles (on an atomic scale) it is impossible to determine precisely both the simultaneous position and momentum of the particle. Any experimental attempt to observe the particle, the simple action of observing, will cause an inherent uncertainty since those photons will interact with the particles. Mathematically, the Heisenberg uncertainty principle states that

$$\Delta x \Delta p \geq \frac{h}{4\pi}$$

where h is Planck's constant.

Spectral Lines

A hot source that emits white light is said to be **incandescent**. Its light, when analyzed with a prism, is broken up into a continuous spectrum of colors. (A **spectroscope** is a device that uses either a prism or a grating to create a spectrum.) If a colored filter is used over the incandescent source, light of only that one color will be emitted, and the spectrum will appear as a continuous band of that color.

If hydrogen gas contained at low pressure in a tube is electrically sparked, the gas will emit a bluish light. If that light is analyzed with a spectroscope, a discrete series of colored lines ranging from red to violet, rather than a continuous band of color, is revealed. When sparked, other chemical elements also show characteristic **emission line spectra** that are useful for chemical analysis. These spectra provided the evidence for Bohr to quantize the energy levels of electrons.

Figure 7.7 illustrates a typical hydrogen spectral series in the visible range of the electromagnetic spectrum. This series is often called the **Balmer series** in honor of Johann Jakob Balmer, a Swiss mathematician who empirically studied the hydrogen spectrum in the late nineteenth century. In Figure 7.7, a hydrogen tube is excited and its light is passed through a narrow slit to provide a point source. This light is then passed through a prism or grating. Many emission lines are produced, but only the first four, designated as H_α (red), H_β, H_γ, and H_δ (violet), are illustrated here. The wavelengths range from about 656 to about 400 nanometers.

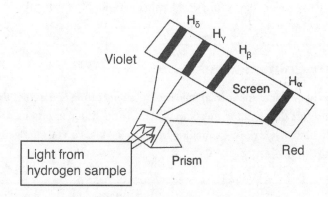

Figure 7.7 Spectroscopy

Special Relativity

In 1905, Einstein published several papers, one of which is now seen as answering the puzzling experimental observation that the speed of light is the same to all observers. Unlike any other relative velocity problem, light's speed does not change when you change reference frames. No matter the various speeds of different observers, they will each measure the exact same speed for any electromagnetic wave in a vacuum, 3×10^8 m/s (c). Rather than begin with the experimental evidence (most of which came after his paper), Einstein began with the theories of electricity and magnetism. When combined, these yield a speed for electromagnetic radiation of 3×10^8 m/s *independent of the speed of the source of the radiation.*

Instead of trying to tinker with the theories governing electricity and magnetism (which were, by that time, pretty well verified), Einstein took the universality of the speed of light as a fundamental property of space itself. He explained that all fields (or, more generally, any object with zero rest mass) propagate through space at this speed. In other words, what we call the speed of light is not just some property of light. It is the speed limit of the universe itself. To address the issue of how different observers could possibly measure the same speed when the observers have different velocities relative to each other, Einstein forces us to look at what it means to take measurements of both distance and time (obviously required when measuring speed). By careful analysis of the fact that two measurements need to be made (ending and starting points of the intervals in space or the intervals in time), he demonstrated that the idea of simultaneity is relative. Two events that may appear simultaneous to one observer will not appear so to another observer as the information about the two events must propagate at the finite speed of light.

This lack of simultaneity on the part of two independent observers led Einstein to conclude that different observers will not agree about intervals of space and time either. As people more commonly say, the two observers will not agree with each other's metersticks and stopwatches. Someone traveling at a different speed from you will appear to have shorter metersticks and slower clocks. If the difference in distance measured by each observer is divided by the difference in time measured by each observer, the values will cancel out. As a result, each observer will come up with the same number of meters traveled per second for electromagnetic radiation. This occurs because the speed of light is finite.

The equations for length contraction and time dilation both depend on the gamma factor, where v is the relative velocity between two observers:

$$\gamma^2 = 1/(1 - v^2/c^2)$$
$$\Delta T_{moving} = \gamma \Delta T_{rest}$$
$$\Delta D_{moving} = \Delta D_{rest}/\gamma$$

As surprising as Einstein's propositions are, they have been experimentally verified many times over. They are used every day by physicists who deal with high-velocity objects.

> **TIP**
>
> Although these equations are unlikely to appear on the AP Physics 2 exam, the concepts of length contraction and time dilation are important ones.

Mass-Energy Equivalence

In 1905, in a follow-up paper to his special theory of relativity published earlier that year, Einstein proposed that mass and energy are different manifestations of the same thing. In other words, neither mass nor energy is conserved by itself. Rather, it is the combination of the two. Mass can be changed into energy and vice versa via his famous relationship:

$$E = mc^2$$

where c is the speed of light. The speed of light is such a large number that everyday transfers of energy do not result in any meaningful change in mass. (However, a cooling coffee mug is indeed losing mass as it radiates away energy!) In low-mass, high-energy situations, the changes in mass due to energy loss or gain become impossible to ignore. For example, in nuclear fusion and fission, the total number of nucleons is the same before and after, yet the total mass of the product nuclei are lower after energy has been released and are higher after energy has been absorbed. In fact, most of the mass of compound objects like atoms, molecules, and people comes from the bending energies within the nuclei of atoms. The individual quarks and electrons by themselves provide a very minor contribution to the total mass of everyday objects.

> **Sample Problem 3**

A hot cup of coffee loses 50 joules of energy as it cools down. By how much has its mass decreased?

✓ **Solution**

$$m = E/c^2 = 50\,\text{J}/(3 \times 10^8\,\text{m/s})^2 = 5.6 \times 10^{-14}\,\text{kg}$$

It is not likely you will notice this loss when you lift the cup!

> **Sample Problem 4**

An electron meets a positron (the antiparticle of an electron, a positively charged electron), and they annihilate each other. How much energy is released?

✓ **Solution**

$$E = mc^2 = (M_{\text{electron}} + M_{\text{positron}})c^2 = (2 \times 9.11 \times 10^{-31})(3 \times 10^8\,\text{m/s})^2 = 1.6 \times 10^{-13}\,\text{J}$$

This value is over 1,000 times greater than the binding energy of electrons in hydrogen atoms!

Atomic Structure and Rutherford's Model

In 1896, French physicist Henri Becquerel was studying the phosphorescent properties of uranium ores. Certain rocks glow after being exposed to sunlight, and many physicists at that time were examining the light emanations from these ores.

One day, while engaged in such a study, Becquerel placed a piece of uranium ore in a drawer on top of some sealed photographic paper. Several days later, he found to his surprise that the paper had been exposed even though no light was incident on it! His conclusion was that radiation emitted by the uranium had penetrated the sealed envelope and exposed the paper.

Subsequent work by Marie and Pierre Curie revealed that this natural radioactivity, as they called it, came from the uranium itself and was a consequence of its instability as an atom. Also, Ernest Rutherford discovered (see Figure 7.8) that the radiation emitted yielded particles that were useful for further atomic studies.

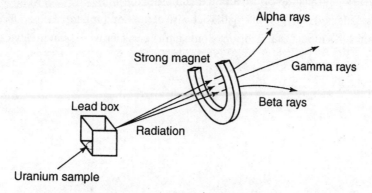

Figure 7.8 Three Types of Radiation

When passed through a magnetic field, the emitted radiation split into three separate parts, called **alpha rays**, **beta rays**, and **gamma rays**. Analysis of their trajectories led to the conclusions that alpha particles, later identified as helium nuclei, are positively charged; beta particles, later identified as electrons, are negatively charged; and gamma rays, later identified to be photons, are electrically neutral.

J. J. Thomson, using the data available at the end of the nineteenth century, devised an early model of the atom based on the fact that all atoms are electrically neutral. After concluding that the atom is neutral overall and yet has electrons also associated with it, Thomson's model proposed that the atom consisted of a relatively large, uniformly distributed, positive mass with negatively charged electrons embedded in it like raisins in a pudding (see Figure 7.9).

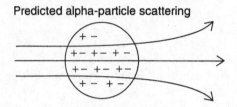

Figure 7.9 Thomson's Model of the Atom

In 1911, Rutherford and his coworkers decided to test Thomson's model using alpha-particle scattering. In the Thomson model, the positive charge was uniformly distributed. Since the negative electrons were embedded throughout the atom as well, the Coulomb force of attraction would weaken the deflecting force on a passing positively charged alpha particle.

Rutherford's procedure was to aim alpha particles at a thin metal foil (he used gold) and observe the scattering pattern on a zinc sulfide screen (Figure 7.10).

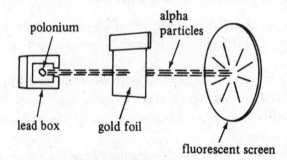

Figure 7.10 Rutherford's Experiment

When the results were analyzed, Rutherford found that most of the alpha particles were not deflected. A few, however, were strongly deflected in hyperbolic paths because of the Coulomb force of repulsion. Since these forces were strong, Thomson's model was incorrect. Some alpha particles were scattered back close to 180 degrees (Figure 7.11).

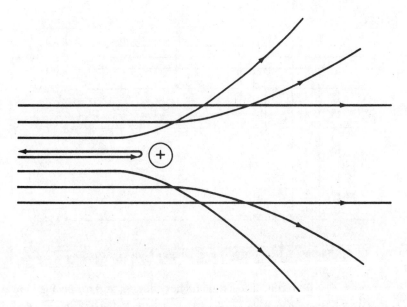

Figure 7.11 Alpha Particle Scattering Trajectories

Rutherford's conclusion was that the atom consisted mostly of empty space and that the nucleus was a very small, densely packed, positive charge. The electrons, he proposed, orbited around the nucleus like planets orbiting the Sun.

Rutherford's model, however, had several major problems, including the fact that it could not account for the appearance of discrete emission line spectra. In Rutherford's model the electrons continuously orbited around the nucleus. This circular, "accelerated" motion should produce a continuous band of electromagnetic radiation, but it did not. Additionally, the predicted orbital loss of energy would cause an atom to disintegrate in a very short time and thus break apart all matter. This phenomenon, too, did not occur.

The Bohr Model

In 1913, Niels Bohr, a Danish physicist, decided to study the hydrogen spectral lines again. Years earlier, Einstein had proposed a quantum theory of light, developing the concept of a photon and completely explaining the photoelectric effect.

Bohr observed that the Balmer formula for the hydrogen spectrum could be viewed in terms of energy differences. He then realized that these energy differences involved a new set of postulates about atomic structure. The main ideas of his theory are these:

1. Electrons do not emit electromagnetic radiation while in a given orbit (energy state).
2. Electrons cannot remain in any arbitrary energy state. They can remain only in discrete or quantized energy states.
3. Electrons emit electromagnetic energy when they make transitions from higher energy states to lower energy states. The lowest energy state is the **ground state**.

According to the Bohr theory, the energy of a photon is equal to the energy difference between two energy states (Figure 7.12).

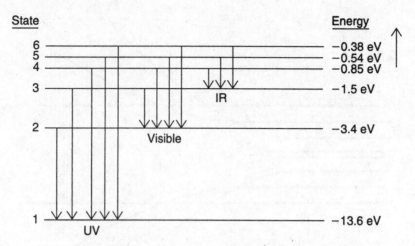

Figure 7.12 Energy-Level Diagram for Hydrogen

The different spectral series arise from transitions from higher energy states to specified lower energy states. The ultraviolet Lyman series occurs when electrons fall to level 1; the visible Balmer series, when electrons fall to level 2; the infrared Paschen series, when electrons fall to level 3. Thus, we can write that the energy, in joules, for an emitted photon is given by

$$f = \frac{E_{\text{initial}} - E_{\text{final}}}{h}$$

The energy-level diagram for hydrogen is shown in Figure 7.12. The energy in a given state represents the amount of energy, called the **ionization energy**, needed to free an electron from the atom. A charged atom is termed an **ion**. Electron ionization energies were confirmed by experiments in 1914.

We designate the levels mathematically as negative so that the ionization level will have a value of zero; the energies carried away by emitted photons are always positive. Even though hydrogen has only one electron, the many atoms of hydrogen in a sample of this gas, coupled with the different probabilities of electrons being in any one particular state, produce the multiple lines that appear in the visible spectrum when the gas is "excited."

> **Sample Problem 5**

What is the wavelength of a photon emitted when an electron makes a transition in a hydrogen atom from level 5 to level 3?

✓ **Solution**

From level 5 to level 3 involves an energy difference of

$$E = (-0.54 \text{ eV}) - (-1.50 \text{ eV}) = 0.96 \text{ eV} = 1.536 \times 10^{-19} \text{ J}$$

Now, the frequency of the photon will be equal to

$$f = \frac{1.536 \times 10^{-19}}{6.63 \times 10^{-34}} = 2.3 \times 10^{14} \text{ Hz}$$

The wavelength of this photon is given by

$$\lambda = \frac{c}{f} = \frac{3 \times 10^8}{2.3 \times 10^{14}} = 1.3 \times 10^{-6} \text{ m}$$

(which is in the infrared part of the electromagnetic spectrum).

It is important to remember that an electron will make a transition from a lower to a higher energy state only when it absorbs a photon equal to a given possible ("allowed") energy difference. This is the nature of the quantization of the atom; the electron can exist only in a specified energy state. To change states, the electron must either absorb or emit a photon whose energy is proportional to Planck's constant and equal to the energy difference between two energy states.

The Heisenberg uncertainty principle, discussed in the section "Matter Waves" earlier in this chapter, introduces the element of "probability" in describing the nature of electron orbits and energy states. This subject is studied in the science of **quantum mechanics**, developed by Erwin Schrödinger and Werner Heisenberg.

Quantum Mechanics and the Electron Cloud Model

The success of the Bohr model was limited to the hydrogen atom and certain hydrogen-like atoms (atoms with one electron in the outermost shell; single-ionized helium is an example). The wave nature of matter led to a problem in predicting the behavior of an electron in a strict Bohr energy state.

In 1926, physicist Erwin Schrödinger developed a new mechanics of the atom, then called **wave mechanics** but later termed **quantum mechanics**. In the wave mechanics of Schrödinger, the position of an electron was determined by a mathematical function called the **wave function**. This wave function was related to the probability of finding an electron in any one energy state. A **wave equation**, called the **Schrödinger equation**, was derived using ideas about time-dependent motion in classical mechanics.

The energy states of Bohr were now replaced by a probability "cloud." In this **cloud model** the electrons are not necessarily limited to specified orbits. A cloud of uncertainty is produced, with the densest regions corresponding to the highest probability of an electron being in a given state.

The use of probability to explain the structure of the atom led to intense debates between the proponents of the new quantum mechanics and one of its fiercest opponents, Albert Einstein. Einstein, who had helped to develop the quantum theory of light twenty years earlier, was not convinced that the universe could be determined by probability. To paraphrase, Einstein said he could not believe that God would play dice with the universe. "God is subtle," he said, "but not malicious."

At about the same time that Schrödinger introduced his wave mechanics, Werner Heisenberg developed his matrix mechanics for the atom. Instead of using a wave analogy, Heisenberg developed a purely abstract, mathematical theory using matrices. Working with Heisenberg, Wolfgang Pauli advanced the so-called **Pauli exclusion principle**, which states that two electrons having the same spin orientations cannot occupy the same quantum state at the same time.

The new **quantum mechanics** was born when the two rival theories were shown to be equivalent. Electron spin was experimentally verified, and the theory proved to be more successful than the older Bohr theory in explaining atomic and molecular structures and led to new ideas about solid-state physics and the subsequent development of lasers.

Nuclear Structure and Stability

Experiments on nuclear masses using a **mass spectrograph**, which deflects the positively charged nuclei into curved paths with different radii, confirmed that the estimate of nuclear mass was too low. The nucleus could not, therefore, be made entirely of positively charged protons.

Physicists theorized that another electrically neutral particle, called the **neutron**, must be present since (a) all of the charge on the nucleus could be accounted for by the presence of the protons, (b) more neutral mass was needed, and (c) a method for countering the repulsive Coulomb forces created by the protons was required.

Thus, the question of nuclear stability depended on the particular structure of the nucleus. The work on radioactive decay by Rutherford and Marie and Pierre Curie confirmed that the radioactive disintegration of an atom is a nuclear phenomenon and is tied to stability as well.

The basic structure of the atom, using a Bohr-Rutherford planetary model, consists of a nucleus with a certain number of protons and neutrons (together called **nucleons**). The number of protons characterizes an element since no two elements have the same number of protons. This **atomic number**, Z, also measures the relative charge on the nucleus. The **mass number**, A, is a measure of the total number of nucleons (protons plus neutrons) in the nucleus. Both of these numbers are whole numbers. However, the actual nuclear mass is not a whole number since it is an average of many atoms of the same element.

Since an atom is electrically neutral, the initial numbers of protons and electrons are the same. Schematically, an atom is designated by a capital letter, possibly followed by one lowercase letter, indicating the element it represents, with the two nuclear numbers A and Z written as a superscript and a subscript, respectively, on the left-hand side. For an unknown element X we write $^A_Z X$. The designation $^1_1 H$ represents the element hydrogen, with one nucleon (a proton) and an atomic number of 1. Some elements have **isotopes**—that is, nuclei with the same number of protons but different numbers of neutrons. For example, the element helium is designated as $^4_2 He$, where the 4 indicates 2 protons and 2 neutrons. An isotope of this atom is helium-3, designated as $^3_2 He$. Some nuclear structures are illustrated in Figure 7.13.

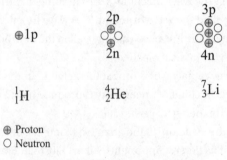

Figure 7.13 Nuclear Structures

A neutron is slightly more massive than a proton. This represents the fact that a neutron has more energy contained within it than does a proton. A proton mass is approximately equal to

$$m_p = 1.673 \times 10^{-27} \text{ kg}$$

and a neutron mass is approximately equal to

$$m_n = 1.675 \times 10^{-27} \text{ kg}$$

Nuclear masses are usually expressed in **atomic mass units** (u). One atomic mass unit is defined to be equal to one-twelfth the mass of a carbon-12 isotope:

$$1 \text{ u} = 1.66 \times 10^{-27} \text{ kg}$$

On this scale, the proton's mass is given as 1.0078 atomic mass units, and the neutron's mass is given as 1.0087 atomic mass units.

If one looks at all stable isotopes of known nuclei (see Figure 7.14), something qualitatively interesting emerges. Experiments show that lighter nuclei have about the same number of protons as neutrons. However, heavier stable nuclei have more neutrons than protons. This suggests that the neutrons must somehow help overcome the repulsive forces in a tightly packed positive nucleus. Also, this **strong nuclear force** must operate within the short distances in an atomic nucleus. Unstable, radioactive nuclei lie off this stability curve. As they decay, the alpha, beta, and other particle emissions **transmute** the original nucleus into a more stable one.

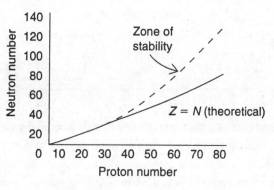

Figure 7.14 Proton-Neutron Plot

REMEMBER

Electric forces repel all protons away from each other, tearing apart the nucleus. Both protons and neutrons are bound by the strong nuclear force, holding the nucleus together. This balancing act of forces gives us both stable and unstable nuclei.

Binding Energy

If we were to determine the total mass of a particular isotope, say $_2^4\text{He}$, based on the number of nucleons, we would arrive at a nucleus that was more massive than in reality. For example, the helium-4 nucleus contains two protons (each with a mass of 1.0078 u) and two neutrons (each with a mass of 1.0087 u). The total predicted mass would be 4.0330 u. Experimentally, however, the mass of the helium-4 nucleus is found to be 4.0026 u. The difference of 0.0304 u, called the **mass defect**, is proportional to the **binding energy** of the nucleus ($E = mc^2$).

Using Einstein's special theory of relativity, we know that mass can be converted to energy. The energy equivalent of 1 atomic mass unit is equal to 931.5 million electron volts (MeV). The binding energy of any nucleus is therefore determined by using the relationship

$$\text{BE (MeV)} = \text{Mass defect} \times 931.5 \text{ MeV/u}$$

In our example, the binding energy of the helium-4 nucleus is

$$\text{BE} = 0.0304 \times 931.5 = 28.3 \text{ MeV}$$

It is useful to think about the average binding energy per nucleon, which is equal to BE/A, where A = the mass number of the element. Thus, for helium-4, where $A = 4$,

$$\text{Avg BE} = \frac{28.3}{4} = 7.08 \text{ MeV per nucleon}$$

In Figure 7.15, a graph of average binding energy per nucleon is plotted against atomic number Z. The graph shows that a maximum point is reached at about $Z = 26$ (iron). The greater the average binding energy, the greater the stability of the nucleus. Thus, since hydrogen isotopes have less stability than helium isotopes, the **fusion** of hydrogen into helium increases the stability of the nucleus through the release of energy. Likewise, the unstable uranium isotopes yield more stable nuclei if they are split (**fission**) into lighter nuclei (also releasing energy).

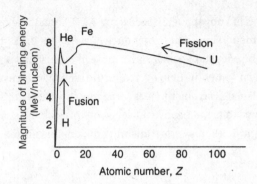

Figure 7.15 Strength of Nuclear Binding Energies

Radioactive Decay

Radioactivity was discovered in 1896 by Henri Becquerel, and the relationship between radioactivity and nuclear stability was studied by Pierre and Marie Curie. Ernest Rutherford pioneered the experiments of nuclear bombardment and introduced the word **transmutation** into the nuclear physics vocabulary. The discovery of alpha- and beta-particle emissions from nuclei led to the use of these particles to probe the structure of the nucleus in the first decades of the twentieth century. Subsequent research showed the existence of positively charged electrons called **positrons** and even negatively charged protons!

The question of how these particles could originate from a nucleus was answered using Einstein's theory of mass-energy equivalence ($E = mc^2$). Schematically, we have the following designations for some of these particles, as outlined in Table 7.1.

Table 7.1 Common Subatomic Particles Beyond Simply Protons, Neutrons, and Electrons, Emitted by Nuclear Decay

Name	Description	Symbol
Alpha particle	+2 charge, 2 protons and 2 neutrons, nucleus of a helium atom	α
Neutrino/antineutrino	Neutral and extremely low mass, only interact via weak nuclear force and gravity	v and $\bar{v}$
Positron	Antielectron: same mass as an electron but positively charged	e^+ or β^+
Photon	Individual quantum of high-frequency gamma radiation	γ

The process by which an unstable nucleus decays into a more stable one is an inherently probabilistic one. As such, no one can predict when any one particular nucleus will undergo this decay. However, large sample sizes of unstable nuclei can be categorized by their half-life. The **half-life**, $t_{1/2}$, is the time it takes, on average, for half of the amount of material present to undergo its radioactive decay. A radioactive substance with a short half-life will decay quickly, and the unstable, original nuclei will be replaced by their stable products relatively quickly. On the other hand, radioactive atoms with a long half-life will only decay very slowly, and thus the radioactive material will be around for a long time.

By defining a decay constant (λ) in terms of the half-life, we can write an expression for the number of particles left (N) from a starting sample of N_0 after a time t as:

$$N = N_0 e^{-\lambda t}$$

TIP

The AP Physics 2 exam will not ask about the following topics: neutron emission, electron capture, the weak nuclear force, or types of neutrinos.

where

$$\lambda = \frac{\ln 2}{t_{1/2}}$$

and which can be rearranged as

$$\ln\left(\frac{N}{N_0}\right) = -\lambda t$$

The type of radioactive decay a given nucleus will undergo is determined by the isotope of the element. There are four pathways for radioactive decay, as summarized in Table 7.2. Note that all conserve the following three quantities: (1) the number of nucleons (neutrons + protons), (2) the total charge, and (3) the number of leptons (electrons + neutrinos).

Table 7.2 Types of Radioactive Decay

Process	Action	Result
Alpha decay	Nucleus ejects an alpha particle	Nucleus atomic number decreases by 2; mass number decreases by 4
Beta-minus decay	Nucleus emits an electron and an antineutrino	One neutron in the nucleus converts into a proton; atomic number increases by 1
Beta-plus decay	Nucleus ejects an antielectron (positron) and a neutrino	One proton in the nucleus converts into a neutron; atomic number decreases by 1
Gamma decay	Nucleus emits a photon	Higher energy state nucleus settles to a lower energy state

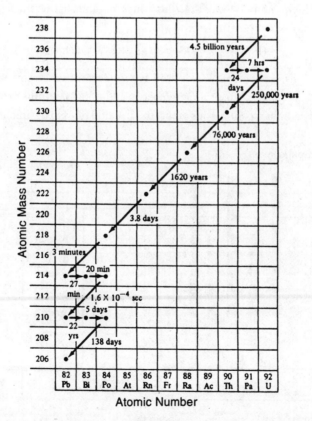

Figure 7.16 Half-Lives and Radioactive Decay Chains of Uranium

When $_{92}^{238}\text{U}$, an isotope of uranium, decays, it emits an alpha particle according to the following equation:

$$_{92}^{238}\text{U} \rightarrow {}_{90}^{234}\text{Th} + {}_{2}^{4}\text{He}$$

When, in turn, the radioactive thorium isotope decays, it emits a beta particle (see Figure 7.16).

$$_{90}^{234}\text{Th} \rightarrow {}_{91}^{234}\text{Pa} + {}_{-1}^{0}e$$

In the first example, both the atomic number and the mass number changed as the uranium transmuted into thorium. In the decay of the thorium nucleus, however, only the atomic number changed. The reason is that in beta decay a neutron in the nucleus is converted into a proton, an electron, energy, and a particle called an **antineutrino**.

In addition to alpha and beta decay, there are several other scenarios for nuclear disintegration. In **electron capture**, an isotope of copper captures an electron into its nucleus and transmutes into an isotope of nickel:

$$_{29}^{64}\text{Cu} + {}_{-1}^{0}e \rightarrow {}_{28}^{64}\text{Ni}$$

In **positron emission**, the same isotope of copper decays by emitting a positron and transmutes into the same isotope of nickel:

$$_{29}^{64}\text{Cu} \rightarrow {}_{+1}^{0}e + {}_{28}^{64}\text{Ni}$$

Note that total charge and mass number are conserved in these reactions.

Fission

Nuclear fission was discovered in the late 1930s by German scientists Otto Hahn and Fritz Strassmann. When $_{92}^{235}\text{U}$, an isotope of uranium, absorbs a slow-moving neutron, the resulting instability will split the nucleus into two smaller (but still radioactive) nuclei and release a large amount of energy. The curve in Figure 7.15 shows that the binding energy per nucleon for uranium is lower than the values for other, lighter nuclei. One possible uranium fission reaction could occur as follows:

$$_{92}^{235}\text{U} + {}_{0}^{1}\text{n} \rightarrow {}_{54}^{140}\text{Xe} + {}_{38}^{94}\text{Sr} + 2\,{}_{0}^{1}\text{n} + 200 \text{ MeV}$$

In controlled fission reactions, water or graphite is used as a **moderator** to slow the neutrons since they cannot be slowed by electromagnetic processes. The production of two additional neutrons means that a chain reaction is possible. In a fission reactor, cadmium control rods adjust the reaction rates to desirable levels. See Figure 7.17.

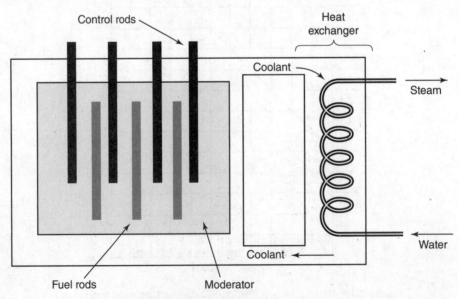

Figure 7.17 Nuclear Reactor

In the production of plutonium as a fissionable material, another isotope of uranium, $^{238}_{92}\text{U}$, is used. First, the uranium absorbs a neutron to form neptunium, a rare element. The neptunium then decays to form plutonium:

$$^{238}_{92}\text{U} + {}^{1}_{0}\text{n} \rightarrow {}^{239}_{92}\text{U} \rightarrow {}^{239}_{93}\text{Np} + {}^{0}_{-1}\text{e}$$

$$^{239}_{93}\text{Np} \rightarrow {}^{239}_{94}\text{Pu} + {}^{0}_{-1}\text{e}$$

Fusion

The binding energy curve (Figure 7.15) indicates that, if hydrogen can be fused into helium, the resulting nucleus will be more stable and energy can be released. In a fusion reaction, isotopes of hydrogen known as deuterium $^{2}_{1}\text{H}$ and tritium $^{3}_{1}\text{H}$ are used. The following equations represent one kind of hydrogen fusion reaction, the kind that powers the Sun:

$$^{1}_{1}\text{H} + {}^{1}_{1}\text{H} \rightarrow {}^{2}_{1}\text{H} + {}^{0}_{1}\text{e} + 0.4 \text{ MeV}$$

$$^{1}_{1}\text{H} + {}^{2}_{1}\text{H} \rightarrow {}^{3}_{2}\text{He} + 5.5 \text{ MeV}$$

$$^{3}_{2}\text{He} + {}^{3}_{2}\text{He} \rightarrow {}^{4}_{2}\text{He} + 2\,{}^{1}_{1}\text{H} + 12.9 \text{ MeV}$$

In this reaction, hydrogen is first fused into deuterium. The deuterium is then fused with hydrogen to form the isotope helium-3. Finally, the helium-3 isotope is fused into the stable element helium.

In a different reaction, the isotope tritium is fused with deuterium to form the stable element helium in one step. This reaction powers the hydrogen (atomic) bomb:

$$^{2}_{1}\text{H} + {}^{3}_{1}\text{H} \rightarrow {}^{4}_{2}\text{He} + {}^{1}_{0}\text{n} + 17.6 \text{ MeV}$$

SUMMARY

- The photoelectric effect concludes that light consists of discrete packets of energy called photons.
- Each photon has an energy proportional to its frequency ($E = hf$).
- In the photoelectric effect, the maximum kinetic energy of the emitted electrons is independent of the intensity (number of photons) and depends directly on the frequency of the incident electromagnetic radiation.
- There is a minimum frequency needed to emit electrons called the threshold frequency.
- The maximum kinetic energy of the electrons can be measured by applying a negative electric potential across their paths called the stopping potential.
- The Compton effect demonstrates that photons have a momentum $p = h/\lambda$.
- Louis de Broglie developed a theory of matter waves in which material particles have a wave-like behavior inversely proportional to their momentum ($\lambda = h/mv$).
- Spectral lines provide an observational demonstration of discrete energy transfers in an atom.
- Ernest Rutherford used alpha particles scattering off of gold foil to show that the atom consists of a small, positively charged nucleus surrounded by negatively charged electrons.
- Niels Bohr developed a theory to explain spectral line emission as well as the predicted loss of electron energy according to classical radiation theory.
- In an atom, according to Bohr, electrons exist in certain discrete energy states. The lowest energy state is called the ground state. Spectral lines are emitted as electrons fall from higher to lower energy states.
- The frequency of emitted photons in a spectral series is proportional to the energy difference between levels ($hf = E_i - E_f$).

- In the quantum mechanical model developed by Erwin Schrödinger, Werner Heisenberg, and others, the atom is surrounded by a probability cloud of electrons due to the uncertainty of locating the exact location and momentum of the electrons (Heisenberg's uncertainty principle).
- The nucleus of an atom consists of neutrons and protons (both called nucleons).
- The number of protons (its charge) is called the atomic number.
- The number of protons and neutrons is called the atomic mass.
- Isotopes of nuclei contain the same number of protons but a different number of neutrons.
- For light nuclei, the ratio of the number of protons to the number of neutrons is almost one-to-one. As atomic numbers increase, there tend to be more neutrons than protons in stable isotopes.
- The binding energy of the nucleus is found from Einstein's equation $E = mc^2$. It is the energy needed to bind the nucleus together and is observed because the actual mass of a nucleus is less than the additive sum of its constituent nucleons.
- Iron has the largest average binding energy per nucleon.
- Radioactive decay is the name given to the process of unstable isotopes (such as uranium) emitting particles and photons.
- Alpha decay is the process by which unstable nuclei emit helium nuclei.
- Beta decay is the process by which unstable nuclei emit electrons or antielectrons.
- Gamma decay is the process by which unstable nuclei emit photons.
- Fission is the process by which a heavy unstable nucleus (like uranium) is split into two smaller daughter nuclei by the absorption of a slow-moving neutron.
- Fusion is the process by which two light nuclei (such as deuterium) are fused into a more stable isotope (such as helium).
- Fusion does not produce radioactive by-products as fission does.
- The Sun and other stars are powered by the fusion of hydrogen into helium. Older stars may fuse helium into other, heavier elements.

Problem-Solving Strategies for Quantum Theory

Remember that the photoelectric effect demonstrates the particle nature of light and is independent of the intensity of the electromagnetic waves used. The number of photons is, however, proportional to the intensity of energy.

Note that in some problems energy is expressed in units of electron volts. This is convenient since the maximum kinetic energy is proportional to the stopping potential. Additionally, wavelengths are sometimes expressed in angstroms, and these values must be converted to meters. Be careful with units, and make sure that SI units are used in equations involving energies and other quantities.

Problem-Solving Strategies for the Nucleus

Solving nuclear equations requires that you remember two rules. First, make sure that you have conservation of charge. That means that all of the atomic numbers, which measure the number of protons (and hence the charge), should add up on both sides. However, to get the actual charge, you must multiply the atomic number by the proton charge in units of coulombs.

The second rule to consider is the conservation of mass. The mass numbers that appear as superscripts should add up. Any missing mass should be accounted for by the appearance of energy (sometimes designated by the letter Q). The mass-energy equivalence is that 1 u is equal to 931.5 MeV. This second rule applies to problems involving binding energy and mass defect. Remember that an actual nucleus has less mass than the sum of its nucleons!

Practice Exercises

1. How many photons are associated with a beam of light having a frequency of 2×10^{16} Hz and a detectable energy of 6.63×10^{-15} J?

   (A) 500
   (B) 135
   (C) 25
   (D) 8

2. A photoelectric experiment reveals a maximum kinetic energy of 2.2 eV for a certain metal. The stopping potential for the emitted electrons is

   (A) 1.2 V.
   (B) 1.75 V.
   (C) 2.2 V.
   (D) 3.5 V.

3. The work function for a certain metal is 3.7 eV. What is the threshold frequency for this metal?

   (A) 7×10^{14} Hz
   (B) 9×10^{14} Hz
   (C) 2×10^{15} Hz
   (D) 5×10^{15} Hz

4. In which kind of waves do the photons have the greatest momentum?

   (A) radio waves
   (B) microwaves
   (C) red light
   (D) X-ray

5. What is the momentum of a photon associated with yellow light that has a wavelength of 5,500 Å?

   (A) 1.2×10^{-17} kg · m/s
   (B) 1.2×10^{-27} kg · m/s
   (C) 1.2×10^{-30} kg · m/s
   (D) 1.2×10^{-37} kg · m/s

6. What is the de Broglie wavelength for a proton ($m = 1.67 \times 10^{-27}$ kg) with a velocity of 6×10^7 m/s?

   (A) 1.5×10^{14} m
   (B) 4.8×10^{-11} m
   (C) 1.5×10^{-14} m
   (D) 6.6×10^{-15} m

7. Which of the following statements about emission line spectra is correct?

   (A) All of the lines are evenly spaced.
   (B) All elements in the same chemical family have the same spectra.
   (C) Only gases emit emission lines.
   (D) All lines result from discrete energy differences.

8. Which electron transition will emit a photon with the greatest frequency?

   (A) $n = 1$ to $n = 4$
   (B) $n = 5$ to $n = 2$
   (C) $n = 3$ to $n = 1$
   (D) $n = 7$ to $n = 3$

9. An electron in the ground state of a hydrogen atom can absorb a photon with any of the following energies except

   (A) 10.2 eV.
   (B) 12.1 eV.
   (C) 12.5 eV.
   (D) 12.75 eV.

10. Which of the following statements about the atom is correct?

    (A) Orbiting electrons can sharply deflect passing alpha particles.
    (B) The nucleus of the atom is electrically neutral.
    (C) The nucleus of the atom deflects alpha particles into parabolic trajectories.
    (D) The nucleus of the atom contains most of the atomic mass.

11. Which of the following statements about the Bohr theory reflect how it differs from classical predictions about the atom?

    I.   An electron can orbit without a net force acting.

    II. An electron can orbit about a nucleus.

    III. An electron can be accelerated without radiating energy.

    IV. An orbiting electron has a quantized angular momentum that is proportional to Planck's constant.

    (A) I and II
    (B) I and III
    (C) III and IV
    (D) II and IV

12. An electron makes a transition from a higher energy state to the ground state in a Bohr atom. As a result of this transition,

    (A) the total energy of the atom is increased.
    (B) the force of attraction on the electron is increased.
    (C) the energy of the ground state is increased.
    (D) the charge on the electron is increased.

13. How many different photon frequencies can be emitted if an electron is in excited state $n = 4$ in a hydrogen atom?

    (A) 1
    (B) 3
    (C) 5
    (D) 6

14. A radioactive atom emits a gamma ray photon. As a result,

    (A) the energy of the nucleus is decreased.
    (B) the charge in the nucleus is decreased.
    (C) the ground-state energy is decreased.
    (D) the force on an orbiting electron is decreased.

15. How many neutrons are in a nucleus of $^{213}_{84}$Po?

    (A) 84
    (B) 129
    (C) 213
    (D) 297

16. The nitrogen isotope $^{13}_{7}$N emits a beta particle as it decays. The new isotope formed is

    (A) $^{14}_{7}$N
    (B) $^{13}_{6}$C
    (C) $^{13}_{8}$O
    (D) $^{14}_{8}$O

17. Which radiation has the greatest ability to penetrate matter?

    (A) X-ray
    (B) alpha particle
    (C) gamma ray
    (D) beta particle

18. In the following nuclear reaction, an isotope of aluminum is bombarded by alpha particles:

    $$^{27}_{13}\text{Al} + {}^{4}_{2}\text{He} \rightarrow {}^{30}_{15}\text{P} + \text{Y}$$

    Quantity Y must be

    (A) a neutron.
    (B) an electron.
    (C) a positron.
    (D) a photon.

19. When a gamma-ray photon is emitted from a radioactive nucleus, which of the following occurs?

    (A) The nucleus goes to an excited state.
    (B) The nucleus goes to a more stable state.
    (C) An electron goes to an excited state.
    (D) The atomic number of the nucleus decreases.

20. The threshold frequency for calcium is 7.7×10^{14} Hz.

    (a) What is the work function, in joules, for calcium?

    (b) If light of wavelength 2.5×10^{-7} m is incident on calcium, what will be the maximum kinetic energy of the emitted electrons?

    (c) What is the stopping potential for the electrons emitted under the conditions in part (b)?

21. Explain why electrons are diffracted through a crystal.

22. Explain why increasing the intensity of electromagnetic radiation on a photoemissive surface does not affect the kinetic energy of the ejected photoelectrons. Why is this kinetic energy referred to as the "maximum kinetic energy"?

23. Why don't we speak of the wavelength nature of large objects such as cars or balls?

24. If a source of light from excited hydrogen gas is moving toward or away from an observer, what changes, if any, are observed in the emission spectral lines?

25. Given the isotope $^{56}_{26}$Fe, which has an actual mass of 55.934939 u:

    (a) Determine the mass defect of the nucleus in atomic mass units.

    (b) Determine the average binding energy per nucleon in units of million electron volts.

26. Why do heavier stable nuclei have more neutrons than protons?

27. Why are the conditions for the fusion of hydrogen into helium favorable inside the core of a star?

28. Explain the radioactive disintegration series of uranium-238 into stable lead-206 in terms of the neutron-proton plot.

29. Albert Einstein, in his special theory of relativity, stated that energy and mass were related by the expression $E = mc^2$. Explain how the concept of binding energy confirms this claim.

Answer Explanations

1. **(A)** If E equals the total detectable energy and E_p equals the energy of a photon, then

$$N = \frac{E}{E_p} = \frac{E}{hf} = \frac{6.63 \times 10^{-15} \text{ J}}{1.326 \times 10^{-17} \text{ J}} = 500$$

2. **(C)** By definition, 1 eV is the amount of energy given to one electron when placed in a potential difference of 1 V. Thus, if $KE_{max} = 2.2$ eV, the stopping potential must be equal to 2.2 V.

3. **(B)** The formula for the work function is $W_0 = hf_0$. The work function must be expressed in units of joules before dividing by Planck's constant:

$$f_0 = \frac{(3.7)(1.6 \times 10^{-19})}{6.63 \times 10^{-34}} = 8.9 \times 10^{14} \text{ Hz}$$

4. **(D)** Since $p = h/\lambda$, the waves with the smallest wavelength will have the greatest photon momentum. Of the four choices, an X-ray photon has the smallest wavelength and thus the greatest momentum.

5. **(B)** The wavelength must first be converted to meters; then we can use $p = h/\lambda$:

$$p = \frac{6.63 \times 10^{-34}}{(5,500)(1 \times 10^{-10})} = 1.2 \times 10^{-27} \text{ kg} \cdot \text{m/s}$$

6. **(D)** The formula for the de Broglie wavelength is $\lambda = h/mv$. Substituting the known values gives

$$\lambda = \frac{6.63 \times 10^{-34}}{(1.67 \times 10^{-27})(6 \times 10^7)} = 6.6 \times 10^{-15} \text{ m}$$

7. **(D)** Spectral emission lines are different for every atom or molecule, and the lines are not evenly spaced. They do, however, arise from energy-difference transitions.

8. **(C)** Transitions to level 1 produce ultraviolet photons that have frequencies greater than those produced by transitions to other levels.

9. **(C)** An electron in the ground state can absorb a photon that has an energy equal to the energy difference between the ground state and any other level. Only 12.5 eV is not such an energy difference.

10. **(D)** "The nucleus of the atom contains most of the atomic mass" is a correct statement.

11. **(C)** The Bohr theory challenged classical theories by stating that an electron can be accelerated without radiating energy (statement III) and that its angular momentum is quantized and proportional to Planck's constant (statement IV).

12. **(B)** As an electron gets closer to the nucleus, the force of attraction on the electron increases.

13. **(D)** There are six possible photon frequencies emitted from state $n = 4$: (1) $n = 4$ to $n = 1$, (2) $n = 4$ to $n = 3$, (3) $n = 4$ to $n = 2$, (4) $n = 3$ to $n = 2$, (5) $n = 3$ to $n = 1$, (6) $n = 2$ to $n = 1$.

14. **(A)** Since photons do not have any rest mass, their emission only releases nuclear energy.

15. **(B)** The number of neutrons is given by

$$N = A - Z = 213 - 84 = 129$$

16. **(C)** The emission of a beta particle involves the transformation of a neutron into a proton and an electron. Thus, the total number of nucleons (Z) remains the same, but the number of protons increases by one. The new isotope is therefore oxygen-13.

17. **(C)** Gamma rays have the greatest energy with the greatest penetration power.

18. **(A)** We must have conservation of charge and mass. The subscripts already add up to 15 on each side, so we are left with a subscript of 0 for Y; the particle is therefore neutral. The superscripts on the right must add up to 31 to balance. Since the missing particle needs a superscript of 1, particle Y is a neutron.

19. **(B)** When radioactive nuclei emit photons, they settle down to a more stable state since no mass has been lost.

20. (a) The work function is given by

$$W_0 = hf_0 = (6.63 \times 10^{-34})(7.7 \times 10^{14}) = 5.1 \times 10^{-19} \text{ J}$$

(b) The frequency associated with a wavelength of 2.5×10^{-7} m is given by

$$f = \frac{c}{\lambda} = \frac{3 \times 10^8}{2.5 \times 10^{-7}} = 1.2 \times 10^{15} \text{ Hz}$$

The energy of the associated photon is equal to

$$E = hf = 7.956 \times 10^{-19} \text{ J}$$

The maximum kinetic energy of the emitted electrons is equal to the difference between this energy and the work function:

$$\text{KE}_{\text{max}} = (7.956 \times 10^{-19}) - (5.1 \times 10^{-19}) = 2.856 \times 10^{-19} \text{ J}$$

(c) This maximum kinetic energy is equal to the product of the electron charge and the stopping potential. Thus:

$$V_0 = \frac{\text{KE}_{\text{max}}}{e} = \frac{2.856 \times 10^{-19}}{1.6 \times 10^{-19}} = 1.785 \text{ V}$$

21. The de Broglie wavelength of electrons is comparable to the spacing between molecules in the lattice structure of a crystal. Thus, the electrons have a wave-like characteristic, and the crystal acts like a diffraction grating.

22. In the quantum theory of light, each photon interacts with one, and only one, electron. Thus, if the energy of a photon is based on its frequency, the intensity of light is just the number of photons. Thus, each electron emerges with the same kinetic energy. This is true for 1 or 1 billion electrons. Therefore, we refer to this kinetic energy as the "maximum kinetic energy."

23. Using the de Broglie formula, it is easy to see that the wavelength of a moving ball or car is much too small to be detected or have any significant wave-like interactions.

24. If the source of the radiation is moving toward or away from an observer, the spectral lines are shifted according to the Doppler effect. In stars, we see redshifts as the stars move away from Earth and blueshifts as they move toward us.

25. (a) The actual mass of iron-56 is 55.934939 u. The total additive mass is determined by noting that this isotope of iron has 26 protons and 30 neutrons. Each proton has a mass of 1.0078 u, and each neutron has a mass of 1.0087 u. Thus, the total constituent mass is

$$26(1.0078) = 26.2028 \text{ u}$$
$$\underline{+\ 30(1.0087) = 30.2610 \text{ u}}$$
$$\text{Total mass} = 56.4638 \text{ u}$$

The mass defect of the nucleus is therefore the difference between this mass and the actual mass:

$$\text{Mass defect} = 56.463800 - 55.934939 = 0.528861 \text{ u}$$

(b) The binding energy is given by

$$\text{BE} = 0.528861 \times 931.5 = 492.63402 \text{ MeV}$$

The average binding energy per nucleon is

$$\text{Avg BE} = \frac{492.63402}{56} = 8.797 \text{ MeV}$$

26. Heavier nuclei have more protons, which increases the Coulomb force of repulsion between them. To maintain stability, extra neutrons are present to provide more particles for the strong nuclear force to act on and overcome this repulsive force (which is indifferent to charge).

27. The density, pressure, and temperature inside the core of a star are sufficient to overcome the natural repulsive tendency of hydrogen atoms to fuse into helium.

28. Radioactive uranium-238 is unstable and off the neutron-proton plot. Through the processes of alpha and beta decay, the emission of neutrons and protons eventually brings the isotopes formed closer to that line of stable nuclei.

29. The release of energy and the missing mass in binding energy are accounted for completely using Einstein's formula. This is true for radioactive decay, fission, and fusion as well as matter-antimatter annihilation.

Practice Tests

ANSWER SHEET
Practice Test 1

1. Ⓐ Ⓑ Ⓒ Ⓓ   11. Ⓐ Ⓑ Ⓒ Ⓓ   21. Ⓐ Ⓑ Ⓒ Ⓓ   31. Ⓐ Ⓑ Ⓒ Ⓓ

2. Ⓐ Ⓑ Ⓒ Ⓓ   12. Ⓐ Ⓑ Ⓒ Ⓓ   22. Ⓐ Ⓑ Ⓒ Ⓓ   32. Ⓐ Ⓑ Ⓒ Ⓓ

3. Ⓐ Ⓑ Ⓒ Ⓓ   13. Ⓐ Ⓑ Ⓒ Ⓓ   23. Ⓐ Ⓑ Ⓒ Ⓓ   33. Ⓐ Ⓑ Ⓒ Ⓓ

4. Ⓐ Ⓑ Ⓒ Ⓓ   14. Ⓐ Ⓑ Ⓒ Ⓓ   24. Ⓐ Ⓑ Ⓒ Ⓓ   34. Ⓐ Ⓑ Ⓒ Ⓓ

5. Ⓐ Ⓑ Ⓒ Ⓓ   15. Ⓐ Ⓑ Ⓒ Ⓓ   25. Ⓐ Ⓑ Ⓒ Ⓓ   35. Ⓐ Ⓑ Ⓒ Ⓓ

6. Ⓐ Ⓑ Ⓒ Ⓓ   16. Ⓐ Ⓑ Ⓒ Ⓓ   26. Ⓐ Ⓑ Ⓒ Ⓓ   36. Ⓐ Ⓑ Ⓒ Ⓓ

7. Ⓐ Ⓑ Ⓒ Ⓓ   17. Ⓐ Ⓑ Ⓒ Ⓓ   27. Ⓐ Ⓑ Ⓒ Ⓓ   37. Ⓐ Ⓑ Ⓒ Ⓓ

8. Ⓐ Ⓑ Ⓒ Ⓓ   18. Ⓐ Ⓑ Ⓒ Ⓓ   28. Ⓐ Ⓑ Ⓒ Ⓓ   38. Ⓐ Ⓑ Ⓒ Ⓓ

9. Ⓐ Ⓑ Ⓒ Ⓓ   19. Ⓐ Ⓑ Ⓒ Ⓓ   29. Ⓐ Ⓑ Ⓒ Ⓓ   39. Ⓐ Ⓑ Ⓒ Ⓓ

10. Ⓐ Ⓑ Ⓒ Ⓓ   20. Ⓐ Ⓑ Ⓒ Ⓓ   30. Ⓐ Ⓑ Ⓒ Ⓓ   40. Ⓐ Ⓑ Ⓒ Ⓓ

Practice Test 1

Section I: Multiple-Choice

TIME: 80 MINUTES

DIRECTIONS: For each question or incomplete statement, select the choice that best answers the question or completes the statement. You may use a calculator and make use of the formula sheet provided in the Appendices.

1. Which materials have the highest heat conductivity?

   (A) gases, because the individual particles move the fastest
   (B) gases, because they are the easiest to ionize
   (C) metals, because they are ductile and malleable
   (D) metals, because they have conduction layers

2. A gas is not able to do work under which of the following circumstances?

   (A) isobaric conditions, because pressure is required for work to be done
   (B) isobaric conditions, because constant force will produce no work
   (C) isochoric conditions, because constant shape implies no changes in energy
   (D) isochoric conditions, because no change in volume implies no displacement

3. As a sound wave enters a medium with a higher wave speed, the sound wave will also experience

   (A) an increase in frequency since the frequency must also rise to match the wave speed.
   (B) an increase in wavelength since the wavelength must also rise to match the wave speed.
   (C) an increase in both frequency and wavelength since their product must equal the wave speed.
   (D) neither an increase in frequency nor an increase in wavelength since the wave was already determined at its source.

4. In which of the following situations would the greatest increase in average atomic velocity be required?

   (A) raising the temperature of one million helium atoms by 100 Kelvin
   (B) raising the temperature of one million neon atoms by 100 Kelvin
   (C) raising the temperature of one million helium atoms by 100 degrees Fahrenheit
   (D) raising the temperature of one million neon atoms by 100 degrees Fahrenheit

GO ON TO THE NEXT PAGE

Questions 5 and 6 are based on the following graphs:

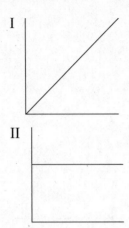

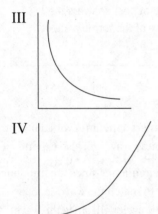

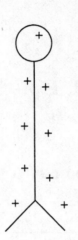

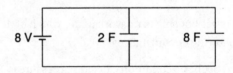

5. Which graph best represents the relation-ship between pressure and volume for an ideal confined gas at constant temperature?

    (A) I
    (B) II
    (C) III
    (D) IV

6. Which graph best represents the relationship between the average kinetic energy of the mole-cules in an ideal gas and its absolute temperature?

    (A) I
    (B) II
    (C) III
    (D) IV

7. The diagram above shows a leaf electroscope that has been charged positively by a negatively charged rod. Which of the following statements is correct?

    (A) The electroscope was charged by conduction.
    (B) The electroscope was charged by contact.
    (C) If the rod is brought closer, protons will move to the top of the electroscope.
    (D) If the rod is brought closer, electrons will be repelled from the top of the electroscope.

Questions 8 and 9 are based on the circuit shown below:

8. What is the maximum charge stored in the 2-farad capacitor?

    (A) 16 C
    (B) 10 C
    (C) 8 C
    (D) 4 C

9. What is the maximum energy stored in the 8-farad capacitor?

   (A) 256 J
   (B) 128 J
   (C) 64 J
   (D) 32 J

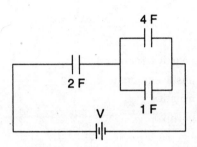

10. What is the equivalent capacitance of the circuit shown above?

    (A) 7/10 F
    (B) 10/7 F
    (C) 14/5 F
    (D) 7 F

11. An electron (charge e, mass m) is trapped in a circular path because of a uniform perpendicular magnetic field B. The velocity of the electron is v, and the radius of the path is r. Which of the following expressions represents the angular velocity ω?

    (A) $(Ber/m)^{1/2}$
    (B) $(Be/rm)^{1/2}$
    (C) Be/m
    (D) $2\pi Be/m$

12. Refer to the following diagram.

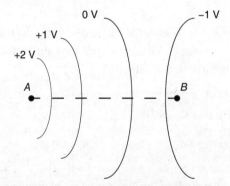

A square wire frame is pulled to the right with a velocity of 4 meters per second across and out of an inward uniform magnetic field of strength 10 teslas. The length of each side of the frame is 0.2 meters. What is the magnitude of the induced motional electromotive force in the wire as it leaves the field?

    (A) 40 V
    (B) 20 V
    (C) 16 V
    (D) 8 V

13. Refer to the figure below.

If a small positive test charge is placed halfway between points A and B, which choice correctly identifies the direction of the force it will experience and its change in electric potential energy as it moves due to that force?

	Direction	Energy
(A)	toward B	Decrease
(B)	toward A	Increase
(C)	toward B	Increase
(D)	toward A	Decrease

GO ON TO THE NEXT PAGE

14. A shadow is formed by a point source of light. Upon closer inspection, the edges of the shadow seem to be diffuse and fuzzy. This phenomenon is probably caused by

(A) dispersion as the different wavelengths of light focus at different points.

(B) refraction as the rays are bent by their contact with the shadow-forming surface.

(C) diffraction as the waves nearest the edge of the shadow-forming surface are sources for waves going into the shadow region.

(D) dispersion as the different wavelengths are traveling at slightly different speeds in the new medium.

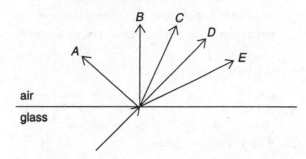

15. A light ray is incident on a glass-air interface as shown above. Which path will the light ray follow after it enters the air?

(A) *A* or *B*

(B) *B* or *C*

(C) *E*

(D) *D*

16. As the angle of incidence for a ray of light passing from glass to air increases, the critical angle of incidence for the glass

(A) increases.

(B) decreases.

(C) increases and then decreases.

(D) remains the same.

17. Which of the following statements about a diverging mirror is correct?

(A) The mirror must be concave in shape.

(B) The images are sometimes larger than the actual objects.

(C) The images are always upright.

(D) The images are sometimes real.

18. Pictured below are two freely hanging balloons; balloon *A* is known to be positive. Which of the following statements best describes balloon *B*?

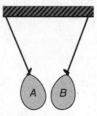

(A) negative or neutral

(B) positive or neutral

(C) negative

(D) positive

19. A student observes that light of a certain wavelength incident upon a thin layer of liquid soap on top of a puddle of oil has no reflection compared to when it reflects off of the oil with no thin layer on top of it. However, the same light seems to be highly reflected from a soap bubble made of the same liquid soap. Which of the following provides the best explanation of her observation?

(A) Spherical surfaces are better at reflecting this wavelength of light compared to flat surfaces.

(B) The light has become so polarized after traveling through the thin layer that it has attenuated to almost nothing when it is on top of the oil.

(C) The reflection off of the inner surface of the soap bubble is not phase shifted, whereas the reflection off of the lower level of the soap film on oil is.

(D) The light from the thin film on the oil is undergoing some interference, whereas the light from the soap bubble is not.

20. In a photoelectric effect experiment, the emitted electrons could be stopped with a stopping potential of 12 volts. What was the maximum kinetic energy of these electrons?

(A) 1.92×10^{-18} J

(B) 12 J

(C) 1.6×10^{-19} eV

(D) 1.92×10^{-18} eV

GO ON TO THE NEXT PAGE

21. As a single photon of light enters a new medium with a higher index of refraction, the photon's energy

    (A) decreases as the wave speed is now lower.
    (B) decreases as its wavelength is now shorter.
    (C) remains constant because its speed is the same.
    (D) remains constant because its frequency remains the same.

22. The capacitor shown below is initially uncharged and the switch is open; after closing the switch, the capacitor is allowed to fully charge. What is the current being drawn from the battery once the capacitor has fully charged?

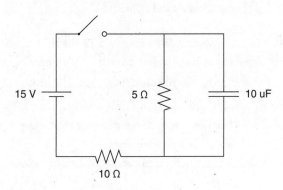

    (A) 0
    (B) 0.5 A
    (C) 1 A
    (D) 3 A

23. A student places an ammeter in parallel with the 12 ohm resistor as shown in the circuit below. Which of the following will be its reading?

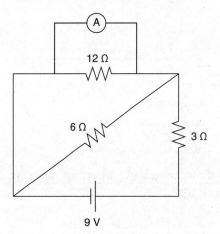

    (A) 3/4 A
    (B) 1 A
    (C) 9/7 A
    (D) 3 A

24. Below is a simulated energy-level diagram for a mythical hydrogen-like atom. Which of the following level transitions will result in the emission of a photon with the highest frequency?

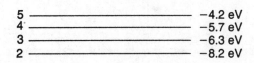

    (A) 1 to 3
    (B) 5 to 2
    (C) 1 to 2
    (D) 2 to 1

25. The mass per nucleon in different elemental nuclei changes because

    (A) the number of nucleons is different for the different elements.
    (B) some elements are radioactive and others are not.
    (C) the binding energies are different in different nuclei.
    (D) the mass per nucleon is the same in all elements.

GO ON TO THE NEXT PAGE

26. Find the magnitude of the magnetic field at point P due to the two parallel lines of current shown.

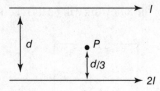

(A) 0

(B) $3\mu_0 I/(2\pi d)$

(C) $9\mu_0 I/(4\pi d)$

(D) $15\mu_0 I/(4\pi d)$

27. In the reaction below, what is the mass number for particle X?

$$^{27}_{13}\text{Al} + ^{4}_{2}\text{He} \rightarrow ^{30}_{15}\text{P} + \text{X}$$

(A) 2

(B) 1

(C) 0

(D) −1

28. Given only two source charges (X and Y) that are creating an electric field as shown below, which of the following choices is incorrect?

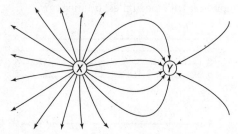

(A) There is a point of zero voltage somewhere between charges X and Y.

(B) From far away, the field will resemble that of a single source equal to $X + Y$.

(C) The magnitude of the electric field just to the left of X is larger than that of the field just to the right of Y.

(D) There is a point of zero electric field somewhere between charges X and Y.

29. Radon gas ($^{222}_{86}\text{Rn}$) is radioactive with a half-life of four days as it undergoes alpha decay. A sample is sealed in an evacuated tube for more than one week. At that time, the presence of a second gas is detected. This gas is most probably

(A) hydrogen.

(B) helium.

(C) nitrogen.

(D) argon.

30. The colors observed in thin films like soap bubbles are caused by

(A) reflection and interference.

(B) refraction and reflection.

(C) diffraction and interference.

(D) polarization and reflection.

31. Which is storing more energy, a 20-micro-farad capacitor charged up by a 6-volt source or a 10-microfarad capacitor charged up by a 12-volt source?

(A) They both store the same amount of energy.

(B) Neither is storing energy.

(C) The 20-microfarad capacitor is storing more energy.

(D) The 10-microfarad capacitor is storing more energy.

32. An electric motor has an effective resistance of 30 ohms, using 4 amperes of current when plugged into a 120-volt outlet. As the motor heats up, its effective resistance increases. Which statement best describes the power consumption of the motor?

(A) It starts off at 480 W and goes down from there as $4^2 R$ as it heats up.

(B) It starts off at 480 W and goes up from there as $4^2 R$ as it heats up.

(C) It starts off at 480 W and goes down from there as $120^2/R$ as it heats up.

(D) It starts off at 480 W and goes up from there as $120^2/R$ as it heats up.

GO ON TO THE NEXT PAGE

33. Which of the following correctly describes the magnetic field near a long, straight wire?

   (A) The field consists of straight lines perpendicular to the wire.
   (B) The field consists of straight lines parallel to the wire.
   (C) The field consists of radial lines originating from the wire.
   (D) The field consists of concentric circles centered on the wire.

34. Electrons are being shot into a uniform magnetic field. The angles of the electrons' velocity vary as they are being shot. A magnetic force will be exerted on all electrons except those that are

   (A) perpendicular to the field.
   (B) parallel to the field.
   (C) at a 45° angle to the field.
   (D) either perpendicular or parallel to the field, depending on the strength of the field.

35. Your friend asks you to explain where the energy comes from in a demo. This demo shows that as a magnet is thrust quickly into a coil of wire, a current is produced, but when a piece of wood is pushed in just as quickly, there is no current produced. How would you explain these findings to your friend?

   (A) The energy is actually the same in both situations.
   (B) The coil of wire used stored magnetic energy to move the charges.
   (C) The energy comes from an application of Faraday's law to the coil.
   (D) The magnet requires more work to be thrust into the coil than the piece of wood.

36. Which of the following statements about the adiabatic expansion of an ideal gas is correct?

   (A) The temperature may change during the expansion.
   (B) The process must be isothermal.
   (C) No change will occur in the internal energy.
   (D) The gas cannot do any work during the expansion.

37. Which of the following processes is not involved in an ideal Carnot cycle?

   (A) isothermal expansion
   (B) isobaric expansion
   (C) adiabatic expansion
   (D) adiabatic compression

38. A charged rod attracts a suspended pith ball. The ball remains in contact with the rod for a few seconds and then is visibly repelled. Which of the following statements must be correct?

   (A) The pith ball is negatively charged at the end of the process.
   (B) The rod is negatively charged.
   (C) The pith ball remained neutral throughout the process.
   (D) The rod has less charge on it at the end of the process than at the beginning.

39. One material is shown to refract light to a greater degree than a second material. This indicates that

   (A) the first material is more opaque than the second.
   (B) the first material has heavier atoms than the second.
   (C) the speed of light is slower in the first material.
   (D) the first material is more magnetic than the second.

PRACTICE TEST 1

40. Two students are arguing about what would happen to the sound they would hear from a stationary police siren while they drove down a cross street, as shown in the diagram.

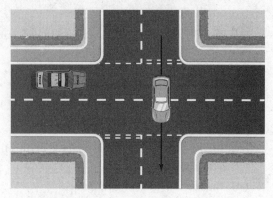

Which of the following arguments is the best physics answer?

(A) "Although the siren will sound loudest as we cross the intersection, the pitch will not change since we are not driving directly toward or away from the siren."

(B) "Since the frequency is set by the source, we never hear a change in pitch due to our own motion, only changes in amplitude, which affects how loudly we perceive the siren."

(C) "Since the police car is not in motion, there will be no Doppler shift and therefore no changes in pitch."

(D) "We will hear a slight increase in pitch as we approach the intersection, as we are partly approaching the source. Then we will hear a slight decrease as we leave the intersection since we are traveling away from the source."

STOP   If there is still time remaining, you may review your answers.

Section II: Free-Response

TIME: 100 MINUTES

DIRECTIONS: Answer all parts of each of the following free-response questions using complete sentences. You may use a calculator and make use of the formula sheet provided in the Appendices.

---

1. (10 points; suggested time: 20–25 minutes)

    A circular conducting loop of radius r is held at rest in the xy-plane, as shown in Figure 1. The resistance of the loop is R. The loop is in a region with a uniform magnetic field of magnitude B_0 that is directed into the page ($-z$-direction). During a time interval t_1, the loop is rotated 90 degrees about the y-axis so that it is perpendicular to the page (the right-hand side of the loop initially moves upward ($+z$), while the left-hand side moves downward ($-z$)).

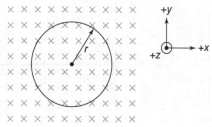

Figure 1

   (a) Express your answers for (a)(i) and (a)(ii) in terms of r, R, B_0, t_1, and physical constants, as appropriate.

   i.   **Derive** an expression for the emf induced in the loop during the interval t_1 while the loop is rotating.

   ii. **Derive** an expression for the total energy dissipated by the loop during the interval t_1.

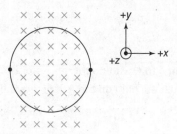

Figure 2

   iii. Complete the following tasks on Figure 2:

   - **Indicate** the direction of the induced current I in the loop during the interval t_1, clearly indicating clockwise or counterclockwise.
   - **Draw** two arrows labeled F to indicate the direction of the magnetic force exerted on the loop (one on the left-hand side and one on the right-hand side) during the interval t_1.

GO ON TO THE NEXT PAGE

(b) Now, instead of rotating the loop, it is held in position while the magnetic field is dialed down to zero over the same time period t_1. **Indicate** whether the induced emf will be different or the same as your answer above:

_____ Induced emf is the same

_____ Induced emf is different

Justify your answer using physics principles.

2. (12 points; suggested time: 25–30 minutes)

An ideal gas expands from points A to C along three possible paths.

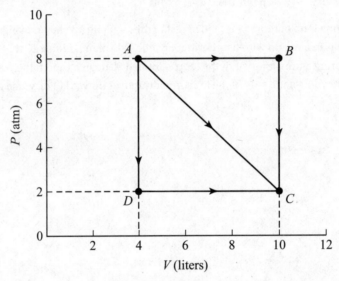

(a) Is it expected that the final temperature at point C be path dependent? **Justify** your answer qualitatively with no calculations.

(b) **Discuss** and **compare** the flow of thermal energy along the three paths (AC, ABC, and ADC). **Indicate** the direction of heat for each pathway (into or out of the gas) and the relative amount of thermal energy involved. **Justify** your answer qualitatively with no calculations.

(c) **Calculate** the work done along the three paths:

   i. ABC

   ii. AC

   iii. ADC

(d) Is there a way to go from A to C with no thermal energy exchanged with the environment? If so, **describe** and **sketch** the path on the P-V diagram below:

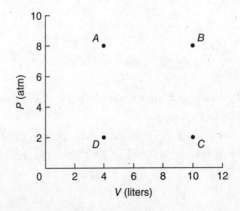

GO ON TO THE NEXT PAGE

(e) **Indicate** whether your answers to parts (b), (c), and (d) above are consistent with your answer to part (a). Briefly **justify** your answer.

3. (10 points; suggested time: 25–30 minutes)

    (a) A laser of unknown wavelength is provided to students along with a screen with two closely spaced slits (1.67 μm far apart) cut in it. Students are also provided with metersticks.

        i.   **Draw** a picture of the experimental setup that the students could use to determine the wavelength of the laser.

        ii.   **Describe** the overall procedure, providing sufficient details and any steps needed to reduce uncertainty.

    (b) **Describe** how the data collected in part (a) could be graphed such that the slope would be the wavelength of the laser.

    (c) In a different experiment, the students are told that the wavelength of the laser is 632.8 nm, but they do not know the spacing between the two slits. The students gather data and place it in the first two columns of the table below.

Angle to the Observed Fringe (degrees)	Number of the Fringe	
0	0 (central, brightest fringe)	
1.6	1	
3.3	2	
4.9	3	
6.6	4	
8.3	5	

        i.   **Indicate** two quantities that could be graphed to yield a straight line, which could then be used to determine the space between the slits. Use the blank column in the table to list any calculated quantities you will graph (other than the data provided).

               Vertical Axis:_____   Horizontal Axis:_____

        ii.   **Plot** the data points for the quantities indicated in part (c)(i) on the graph provided. Clearly scale and label all axes, as appropriate.

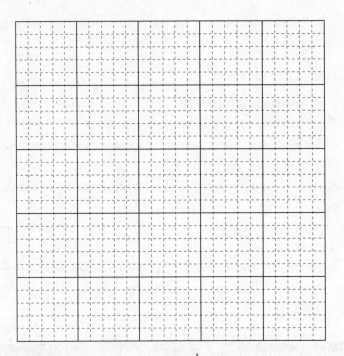

(d) **Draw** a best-fit line on your graph from part (c)(ii), and **calculate** the spacing between the slits.

4.   (8 points; suggested time: 15–20 minutes)

Given the following information:

$$\text{Proton mass} = 1.0078 \text{ u}$$

$$\text{Neutron mass} = 1.0087 \text{ u}$$

$$\text{Mass of } {}^{226}_{88}\text{Ra} = 226.0244 \text{ u}$$

(a) **Determine** the mass defect for this isotope of radium.

(b) What does this mass defect represent? **Describe** both qualitatively and quantitatively.

(c) If radium-88 naturally undergoes alpha decay, **determine** the nuclear reaction for this process. Be sure to show any energy (Q) required or released by this process. If a new element is formed and you are unsure of its symbol, you may use an X to represent that new element. Use the same isotope notation as that given in the information above.

(d) **Compare** the total mass defects on the reactant side of the reaction in part (c) to the mass defect found in part (a). Select one of the three options below, and **justify** your choice qualitatively, without using equations.

_____ The reactants have the same mass defect.

_____ The reactants have a larger mass defect.

_____ The reactants have a smaller mass defect.

Answer Explanations

Section I

1. **(D)** Metal atoms are closely packed in a lattice formation. They also have free conduction layer electrons. Both of these factors contribute to a metal's efficiency at transferring heat.

2. **(D)** Work requires displacement along with the force provided by the pressure of the gas. Therefore, isochoric processes (no volume changes) involve no work.

3. **(B)** Wave speed is determined by the medium. Wave frequency is determined by the source. It is the wavelength in the wave equation that must change in order to satisfy the other two quantities:

$$\text{wavelength} = \frac{\text{wave speed}}{\text{frequency}}$$

4. **(A)** 100 Kelvin is a larger temperature difference than 100 degrees Fahrenheit since each Kelvin is 9/5 of degrees Fahrenheit. Since the energy required is the same and the number of particles is the same, the average kinetic energy increase is the same. The lighter atom (helium) will increase its speed more for the same increase in kinetic energy.

5. **(C)** Since $PV = nRT =$ constant, P and V must have an inverse relationship.

6. **(A)** Temperature is proportional to kinetic energy.

7. **(D)** The electroscope was charged by induction since it is oppositely charged. Bringing the negative charges closer will drive electrons down toward the leaves as protons are not free to move.

8. **(A)** Each capacitor receives the full 8 V from the battery because the capacitors are in series:

$$C = Q/\Delta V$$
$$2\,\text{F} = Q/8\,\text{V}$$
$$Q = 16\,\text{C}$$

9. **(A)** $\frac{1}{2}CV^2 = \frac{1}{2}(8)(8)^2 = 256\,\text{J}$

10. **(B)** The capacitors with 4 F and 1 F are in parallel. Therefore, they are equivalent to one 5 F capacitor. This 5 F and the 2 F are in series:

$$1/5 + 1/2 = 7/10$$
$$10/7\,\text{F}$$

11. **(C)** Magnetic force causing centripetal acceleration:

$$evB = mv^2/r$$

Solve for v:

$$v = eBr/m$$

This tangential v is equal to ωr:

$$\omega = v/r = (eB/m)$$

12. **(D)** emf $= Blv = (10)(0.2)(4) = 8\,\text{V}$

13. **(A)** Positive charges move from higher voltage to lower voltage. (Recall that electric fields point in this direction and are perpendicular to the equipotential lines shown.) They do this to lower their potential energy:

$$\Delta U_E = q\Delta V = (\text{positive charge})(\text{negative change in voltage})$$

14. **(C)** Diffraction is the bending of wave fronts around obstacles or through openings. In this case, diffraction occurs around the edges of the object causing the shadow. This will cause the edge of the shadow to be slightly less than fully dark at the edges.

15. **(C)** When moving from glass to air, light will speed up. As the light speeds up, it will pivot away from the normal.

16. **(D)** The critical angle is set by the ratio of the indices of refraction.

17. **(C)** Divergent mirrors are convex and produce virtual images that are upright and reduced.

18. **(A)** Since the balloons are attracted, based on the figure, balloon B could simply be negatively charged (opposites attract). On the other hand, balloon B might be neutral but experience some induced charge separation due to the nearby presence of the positive share of balloon A. In this case, the closer side of balloon B will be negative and the farther side will be positive, leading to a net attractive force (static cling).

19. **(C)** In both cases, the light coming into the student's eyes is an interference of the light reflecting off of the top of the thin film and from the bottom of the thin film. When on oil, both reflection events are from a faster to a slower medium and thus both are phase shifted. When in air, only the first reflection is phase shifted as the inner boundary of soap-to-air is a transition from a slower medium to a faster one. What was a destructive interference pattern for the soap-on-oil case thus becomes a constructive one for the soap-in-air case since they are the same thickness (likely 1/4 of a wavelength).

20. **(A)** The stopping potential is how the KE of the emitted electrons is measured:

$$1 \text{ electron} \times 12\,\text{V} = 12\,\text{eV} = 1.92 \times 10^{-18}\,\text{J}$$

21. **(D)** The energy of a photon is associated with its frequency:

$$hf = E$$

Entering a new medium may affect wave speed and wavelength but does not affect frequency.

22. **(C)** Once the capacitor is fully charged, it will play the role of an open switch in the circuit (no current flow in its branch). Therefore, the two remaining resistors are in series for the circuit. This gives the battery an effective resistance of $(5 + 10)$, which equals 15 ohms.

$$I = V/R = 15\,\text{V}/15\,\Omega = 1\,\text{A}$$

23. **(D)** Since the ammeter has an internal resistance of zero, the resistors it is in parallel with (it is also in parallel with the diagonal 6-ohm resistor) are effectively shorted, leaving an effective resistance of $3\,\Omega$ for the battery. Since all of this current will pass through the ammeter, the reading will be:

$$I = V/R = 9\,\text{V}/3\,\Omega = 3\,\text{A}$$

24. **(D)** Photon emission occurs when electrons transition to a lower level. The largest difference in energy emits the highest frequency:

$$E = hf$$

25. **(C)** Because of $E = mc^2$, differences in energy cause differences in mass for the nucleus. More tightly bound nucleons have lower mass-per-nucleon values because of their high negative binding energy.

26. **(C)** First note that, by the right-hand rule, the top current will create a field into the page while the bottom current will create one out of the page. Since they are in opposite directions, we must take the difference in the magnetic fields created by each wire at point P:

Top wire: $\mu_0 I_s/(2\pi r) = \mu_0 I/(2\pi(2d/3))$
$$= 3\mu_0 I/(4\pi d)$$

Bottom wire: $\mu_0 I_s/(2\pi r) = \mu_0(2I)/(2\pi(d/3))$
$$= 6\mu_0 I/(2\pi d)$$

Difference: $6\mu_0 I/(2\pi d) - 3\mu_0 I/(4\pi d) = 9\mu_0 I/(4\pi d)$

27. **(B)** The total mass number (upper number) on each side of the reaction must be equal.

28. **(D)** Since source X must be positively charged and source Y must be negatively charged (by virtue of the direction of the electric field lines), there must be a point in between them where the voltage contributions from each cancel out. Since there are more field lines coming from X than from Y, X must have the larger magnitude of charge. Therefore, from far away, the net positive charge of $X + Y$ is the source for the remaining electric field lines. Since the field lines are more dense on the left side of the diagram than on the right, the field is stronger on the left. Finally, since the field line contributions from both charges are pointed to the right in between X and Y, there is no point of cancellation for the electric field in this area.

29. **(B)** Radon undergoes alpha decay. Since alpha particles are the nuclei of helium, the new gas is most likely helium because these alpha particles have picked up some electrons.

30. **(A)** The interference between the ray reflecting off of the top boundary and another ray bouncing off of the lower boundary either produces some bright colors (constructive interference) or does not produce certain colors (destructive interference).

31. **(D)** Compare $\frac{1}{2}C\Delta V^2$ for each capacitor:

$$\frac{1}{2}(20)(6)^2 = 360 \text{ microjoules}$$

$$\frac{1}{2}(10)(12)^2 = 720 \text{ microjoules}$$

32. **(C)**

$$P = I\,\Delta V = (4\,\text{A})(IR) = (4\,\text{A})(4\,\text{A})(30\text{ ohms})$$
$$= 480\,\text{W}$$

As the motor heats up, the effective resistance increases so that the motor draws less current at the same voltage. Therefore, power decreases. You can't use 4^2R since the 4 amps will be changing.

33. **(D)** Magnetic field lines loop around the moving charges.

34. **(B)** There is no magnetic force when the field and the moving charges are in the same direction.

35. **(D)** When creating an emf in the coil, the magnet will encounter a magnetic force against its motion from the induced charge movement. This will require additional force to push the magnet into the coil, which requires more work. This additional work is the energy needed to create the current.

36. **(A)** Adiabatic means no heat transfer. However, work may change the internal energy, which, in turn, may change the temperature.

37. **(B)** A Carnot cycle is an alternating cycle of adiabatic and isothermal operations.

38. **(D)** For attraction to take place, the pith ball must either have a charge opposite that of the rod or be neutral and undergoing induced charge separation. There is no way to tell what this initial state is without additional information. However, since the pith ball repels at the end of the process, the original charge on the rod must have been partially transferred to the pith ball such that the rod and the pith ball now have the same charge (like charges repel). Because the rod has lost some of its original charge, it is less charged in the end than it was in the beginning.

39. **(C)** Greater refraction indicates a larger index of refraction. Larger indices of refraction are for slower speeds of electromagnetic radiation.

40. **(D)** Although choice (A) is correct insofar as the amplitude (loudness) is concerned, there will also be a change in pitch. This change is caused by the relative motion of the observed and the source changing from toward each other (higher frequency) to being away from each other (lower frequency). This is best captured in choice (D).

Section II

1. (a) i.

$$|\varepsilon| = \left|\frac{\Delta \Phi_B}{\Delta t}\right| = B_0 \Delta A / \Delta t = B_0(0 - \pi r^2)/t_1 = \pi r^2 B_0/t_1$$

[1 point for providing a general expression for the change of magnetic flux since the area is changing (as the loop rotates, the area capturing magnetic field lines is decreasing zero), and 2 points for providing a correct simplified expression in terms of the given variables]

ii.

$$\text{Energy} = Pt = (IV)t = (V^2/R)t = (\pi r^2 B_0/t_1)^2 t_1/R = \pi^2 r^4 B_0^2/(Rt_1)$$

[1 point for providing a general expression for energy in terms of electrical power, and 2 points for substituting in the expression for voltage (emf) and simplifying]

iii. A decreasing flux into the page as the loop rotates will induce a current to flow, which creates a magnetic flux into the page (attempting to keep the flux the same). By the right-hand rule, this is a clockwise current.

[1 point for stating that the direction is clockwise]

Use the right-hand rule: the clockwise current will experience an outward force due to the downward magnetic field:

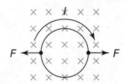

[1 point for both directions of magnetic force]

(b) As the field dissipates, this causes a decrease in the magnetic flux to zero in the same amount of time as before. Therefore, all results will be the same as rotating the loop 90 degrees in the same time.

[1 point for indicating that the induced emf is the same, and 1 point for providing justification]

2. (a) No, the temperature is a state function of pressure and volume only. So the temperature at point C is not path dependent.

[1 point for stating that point C is not path dependent, and 1 point for providing appropriate justification]

(b)

$$\Delta U = Q + W$$

For an ideal gas, the change in internal energy (ΔU) is fixed by the change in the temperature. So the difference in thermal energy exchange (Q) will be dictated by the amount of work done on each path. [1 point]

Without doing a calculation, we can visually tell which path is doing more work by examining the areas under the pathways. The product PV is lower at point C than at point A. Therefore, we can conclude that the temperature is lower at point C (ideal gas law: $PV = nRT$). Since the temperature is decreasing, ΔU is negative, while all the work done by the gas represents a loss of energy as well. The difference between the given pathway and an adiabatic expansion represents the amount of thermal energy exchanged with the environment (Q). [1 point]

Path AC represents moderate work done by the gas, requiring some thermal energy to be added to the gas during the process but not as much as path ABC, which represents the most work done by the gas. [1 point]

Since pathway ABC is so far from adiabatic, it will require significant thermal energy to be added to the gas. [1 point]

Finally, path ADC is the least amount of work done and the only pathway to require thermal energy to be given to the environment from the gas. [1 point]

(c) i.   The work done is going to be equal to the area under each segment of the P-V graph. Also, recall that 1 atm = 101 kPa = 101,000 N/m^2 and 1 L = 0.001 m^3:

$$Pa \cdot m^3 = J$$

Along path ABC, no work is done from $B \to C$ because there is no change in volume. Thus, the area is only under $AB = (8\ \text{atm})(6\ \text{L}) = 4{,}848\ \text{J}$. [1 point]

ii. Along path AC, the total area is equal to:

$$\tfrac{1}{2}(6\ \text{atm})(6\ \text{L}) + (2\ \text{atm})(6\ \text{L}) = 3{,}030\ \text{J}$$

[1 point]

iii. Along path ADC, no work is done from $A \to D$ because there is no change in volume. Thus, the area is only under DC:

$$(2\ \text{atm})(6\ \text{L}) = 1{,}212\ \text{J}$$

[1 point]

(d) Yes, if the expansion is adiabatic, you can go from A to C with no thermal energy exchanged. This would connect point A and point C with a steep hyperbola. In this case, the work done is solely responsible for the change in internal energy as no heat is involved.

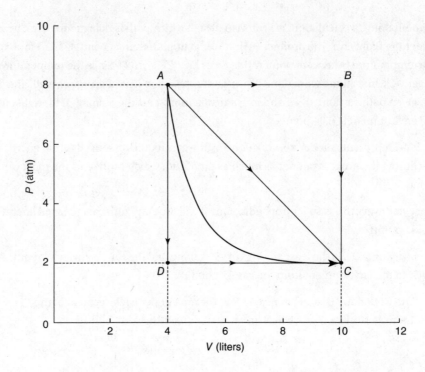

[1 point]

(e) Yes, the answers are all consistent as heat transfer and work done are adding up in all cases to keep the internal energy of the final state the same, no matter the path. [1 point]

3. (a) i.

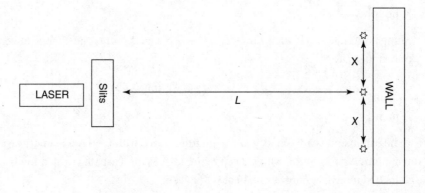

[1 point]

ii. Shine the laser light through the two slits in such a way that the slits are far away from a wall of the room. Note where the projected points of light are located on the wall. You should see the primary bright dot directly in front of the laser and one (or more) less bright spots to either side. The screen with the slits may need to be rotated in order to bring all the bright spots in line (at the same height). Measure the distance from the slits to the wall (L) and the distances between the dots on the wall (X). [1 point]

(b) Using the two-slit diffraction pattern formula:

$$d \sin \theta = m\lambda$$

Use small-angle approximation:

$$d(x/L) = m\lambda$$

where x is the average distance between bright spots, L is the distance between the screen and the wall, m is the number of fringes from the central maximum, and d is the distance between the two slits. [1 point]

By collecting the variables on the left side of the small angle formula above, if one graphs "$d(x/L)$" on the y-axis and "m" on the x-axis, the slope will be the wavelength. [1 point]

(c) i. You could indicate a graph of either (1) "m" on the vertical axis and "$\sin \theta$" on the horizontal axis, giving a slope of d/λ or (2) "$m\lambda$" on the vertical axis and "$\sin \theta$" on the horizontal axis, giving a slope of d.

[1 point for either option]

Angle to the Observed Fringe (degrees)	Number of the Fringe	Sine (angle)
0	0 (central, brightest fringe)	0
1.6	1	0.028
3.3	2	0.058
4.9	3	0.085
6.6	4	0.11
8.3	5	0.14

[1 point for filling in the table with one or more of the calculated values in the third column]

ii.

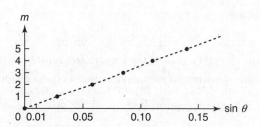

[2 points for correctly plotting the graph, including labels]

(d)

$$\text{Slope} = 3/0.008 = 37.5$$

$$\text{Slope} = d/\lambda$$

$$d = 37.5(\lambda) = (37.5)(632.8 \text{ nm}) = 2.37 \times 10^{-5} \text{ m}$$

[1 point for drawing the line of best fit on the graph from part (c)(ii), and 1 point for finding the slope and determining d]

4. (a) To find the mass defect, we must first find the total constituent mass:

$$88 \text{ protons: } m = (88)(1.0078) = 88.6864 \text{ u}$$
$$138 \text{ neutrons: } m = (138)(1.0087) = 139.2006 \text{ u}$$
$$M(\text{total}) = 227.8870 \text{ u}$$

Then we have:

$$\text{Mass defect} = 227.8870 \text{ u} - 226.0244 \text{ u} = 1.8626 \text{ u}$$

[2 points]

(b) The mass defect represents the additional binding energy holding the nucleus together.

[1 point]

This energy can be found by using Einstein's equation $E = mc^2$. Using the conversion factor from the Table of Information for AP Physics 2 (see the appendices).

$$1 \text{ u} = 931 \text{ MeV}/c^2 \text{ and } E = mc^2$$

The total binding energy (BE) is

$$\text{BE} = (931 \text{MeV/u})(1.8626 \text{ u}) = 1{,}734.0 \text{ MeV}$$

[1 point]

The binding energy per nucleon is:

$$\text{BE/nucleon} = 1{,}734.0/226 = 7.67 \text{ MeV}$$

[1 point]

(c)

or

[1 point for either answer]

(d) The reactants have a larger mass defect. Since this reaction is energy releasing, energy has left the nuclei. This energy has come from the mass, so the mass defect must be increased. The nucleons on the reactant side have greater binding energy on average than those on the product side.

[1 point for selecting the correct option, and 1 point for providing an appropriate explanation]

Test Analysis

Section I: Multiple-Choice

Number correct (out of 40) = $\dfrac{}{\text{Multiple-Choice Score}}$

Section II: Free-Response

Partial credit is awarded for any correct responses within an individual free-response question.

Question 1 = $\dfrac{}{\text{(out of 10)}}$

Question 2 = $\dfrac{}{\text{(out of 12)}}$

Question 3 = $\dfrac{}{\text{(out of 10)}}$

Question 4 = $\dfrac{}{\text{(out of 8)}}$

Total (out of 40) = —————————

Final Score

$$\dfrac{}{\text{Multiple-Choice Score}} + \dfrac{}{\text{Free-Response Score}} = \dfrac{}{\text{Total (out of 80)}}$$

Final Score Range*

Final Score Range	AP Score
60–80	5
44–59	4
32–43	3
20–31	2
0–19	1

*Note: Actual score ranges vary from year to year and are determined by the College Board each year. Thus, the ranges shown are approximate.

ANSWER SHEET
Practice Test 2

1. Ⓐ Ⓑ Ⓒ Ⓓ   11. Ⓐ Ⓑ Ⓒ Ⓓ   21. Ⓐ Ⓑ Ⓒ Ⓓ   31. Ⓐ Ⓑ Ⓒ Ⓓ
2. Ⓐ Ⓑ Ⓒ Ⓓ   12. Ⓐ Ⓑ Ⓒ Ⓓ   22. Ⓐ Ⓑ Ⓒ Ⓓ   32. Ⓐ Ⓑ Ⓒ Ⓓ
3. Ⓐ Ⓑ Ⓒ Ⓓ   13. Ⓐ Ⓑ Ⓒ Ⓓ   23. Ⓐ Ⓑ Ⓒ Ⓓ   33. Ⓐ Ⓑ Ⓒ Ⓓ
4. Ⓐ Ⓑ Ⓒ Ⓓ   14. Ⓐ Ⓑ Ⓒ Ⓓ   24. Ⓐ Ⓑ Ⓒ Ⓓ   34. Ⓐ Ⓑ Ⓒ Ⓓ
5. Ⓐ Ⓑ Ⓒ Ⓓ   15. Ⓐ Ⓑ Ⓒ Ⓓ   25. Ⓐ Ⓑ Ⓒ Ⓓ   35. Ⓐ Ⓑ Ⓒ Ⓓ
6. Ⓐ Ⓑ Ⓒ Ⓓ   16. Ⓐ Ⓑ Ⓒ Ⓓ   26. Ⓐ Ⓑ Ⓒ Ⓓ   36. Ⓐ Ⓑ Ⓒ Ⓓ
7. Ⓐ Ⓑ Ⓒ Ⓓ   17. Ⓐ Ⓑ Ⓒ Ⓓ   27. Ⓐ Ⓑ Ⓒ Ⓓ   37. Ⓐ Ⓑ Ⓒ Ⓓ
8. Ⓐ Ⓑ Ⓒ Ⓓ   18. Ⓐ Ⓑ Ⓒ Ⓓ   28. Ⓐ Ⓑ Ⓒ Ⓓ   38. Ⓐ Ⓑ Ⓒ Ⓓ
9. Ⓐ Ⓑ Ⓒ Ⓓ   19. Ⓐ Ⓑ Ⓒ Ⓓ   29. Ⓐ Ⓑ Ⓒ Ⓓ   39. Ⓐ Ⓑ Ⓒ Ⓓ
10. Ⓐ Ⓑ Ⓒ Ⓓ   20. Ⓐ Ⓑ Ⓒ Ⓓ   30. Ⓐ Ⓑ Ⓒ Ⓓ   40. Ⓐ Ⓑ Ⓒ Ⓓ

Practice Test 2

Section I: Multiple-Choice

TIME: 80 MINUTES

DIRECTIONS: For each question or incomplete statement, select the choice that best answers the question or completes the statement. You may use a calculator and make use of the formula sheet provided in the Appendices.

---

1. In a photoelectric effect experiment, increasing the intensity of the incident electromagnetic radiation will

   (A) increase only the number of emitted electrons.
   (B) have no effect on any aspect of the experiment.
   (C) increase the maximum kinetic energy of the emitted electrons.
   (D) increase the stopping potential of the experiment.

2. Which element will not release energy via either fusion or fission?

   (A) hydrogen
   (B) iron
   (C) helium
   (D) neon

3. Photons can scatter electrons, and the trajectories of the electrons will obey the law of conservation of momentum. These facts are observed in the

   (A) Rutherford scattering experiment.
   (B) Michelson-Morley experiment.
   (C) Compton effect.
   (D) Doppler effect.

4. Look at the following reaction:

   $$^{6}_{3}\text{Li} + \text{X} \rightarrow ^{7}_{4}\text{Be} + ^{1}_{0}\text{n}$$

   What is element X?

   (A) $^{1}_{1}\text{H}$
   (B) $^{3}_{1}\text{H}$
   (C) $^{4}_{2}\text{He}$
   (D) $^{2}_{1}\text{H}$

5. As hydrogen is fused into helium, the stability of the new nuclei, compared with that of hydrogen, will be

   (A) greater because of greater binding energy per nucleon.
   (B) less because of greater binding energy per nucleon.
   (C) greater because of smaller binding energy per nucleon.
   (D) less because of smaller binding energy per nucleon.

6. A ray of light is incident from a slower medium (smaller than the critical angle) to one that is faster. The refracted ray will

    (A) not refract at all.
    (B) refract toward the normal.
    (C) refract away from the normal.
    (D) "refract" at 90 degrees such that it does not actually enter the faster medium.

7. Which of the following expressions has the same SI units as PV?

    (A) mv
    (B) mc^2
    (C) Q/T
    (D) gh

8. Heat conduction through a slab of a certain material depends on all of the following EXCEPT

    (A) area.
    (B) temperature difference between the face of the materials.
    (C) thickness of the material.
    (D) specific heat capacity of the material.

Questions 9–11 are based on the following information:

An RC circuit is wired such that a 10 μF capacitor is connected in series with a 5 Ω resistor and a 6 V battery.

9. What will be the total charge that flows into one side of the capacitor?

    (A) 50 C
    (B) 2 C
    (C) 5×10^{-4} C
    (D) 6×10^{-5} C

10. Which of the following graphs best represents the increase of charge in the capacitor while attached to a constant voltage source?

    (A)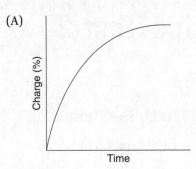

    (B)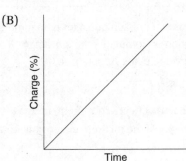

    (C)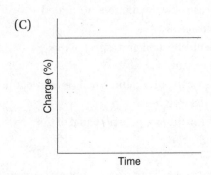

    (D)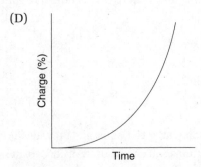

11. What is the value of the total energy stored in the capacitor?

    (A) 1.8×10^{-4} J
    (B) 6×10^{-5} J
    (C) 3×10^{-5} J
    (D) 2×10^{-5} J

GO ON TO THE NEXT PAGE

12. Anti-reflecting coatings on objects are often made to be one-fourth of a wavelength thick. These coatings reduce some reflected intensities primarily because of

    (A) reflection and refraction.
    (B) reflection and interference.
    (C) reflection and polarization.
    (D) dispersion and refraction.

13. The time between maximum electric field strength magnitudes at a fixed location in a vacuum as an electromagnetic wave with a wavelength of 6 meters passes by is

    (A) 5×10^7 s
    (B) 2.5×10^{-7} s
    (C) 2×10^{-8} s
    (D) 1×10^{-8} s

14. Which of the following is the best reason for the observation that a household battery delivers less voltage as you attach it to a lower resistance load?

    (A) The internal resistance of the battery is using more voltage inside the battery.
    (B) The battery is delivering constant power, so the greater the current, the lower the voltage.
    (C) Ohm's law dictates that the lower resistance will draw more current. Current is inversely related to voltage.
    (D) This particular lower resistance load must contain a transformer.

Questions 15–19 refer to the following information:

Light with a wavelength of 5.4×10^{-7} meters falls on two slits that are separated by a distance of 1.0×10^{-3} meters. An interference pattern forms on a screen 2 meters away.

15. What is the energy of each photon of this light?

    (A) 3×10^8 J
    (B) 4×10^{8} J
    (C) 4×10^{-19} J
    (D) 3×10^{-33} J

16. What is the approximate distance between the central bright band and the next bright band?

    (A) 1.08×10^{-3} m
    (B) 2.7×10^{-3} m
    (C) 3.6×10^{-2} m
    (D) 2 m

17. The dark and bright interference pattern observed on the screen is caused by

    (A) reflection and refraction.
    (B) refraction and diffraction.
    (C) polarization and interference.
    (D) diffraction and interference.

18. If the distance between the slits and the screen is increased, the pattern spacing will

    (A) increase.
    (B) decrease.
    (C) increase and then decrease.
    (D) remain the same.

19. If light with a shorter wavelength is used in an experiment, the pattern separation will

    (A) increase.
    (B) decrease.
    (C) increase and then decrease.
    (D) remain the same.

20. As light enters a slower medium (index of refraction n_2) from a faster one (index of refraction n_1), its wavelength is observed to be cut down to 1/3 of its previous value. What is the ratio of n_2/n_1?

    (A) 3.0
    (B) 1.0
    (C) 0.33
    (D) The frequency and/or wavelength must be specified.

21. What is the critical angle for light within a transparent material ($n = 2.0$) when immersed in water ($n = 1.3$)?

    (A) 30°
    (B) 41°
    (C) 60°
    (D) No critical angle in this situation

GO ON TO THE NEXT PAGE

22. A doubly ionized helium atom is accelerated by a potential difference of 400 V. The maximum kinetic energy of the ion is equal to

    (A) 200 eV
    (B) 400 eV
    (C) 800 eV
    (D) 1,600 eV

23. A neutral body is rubbed and becomes positively charged by

    (A) gaining protons.
    (B) losing electrons.
    (C) gaining electrons.
    (D) losing protons.

24. Inside of a complex circuit is a switch that controls a light. When the switch is open, the light is on. When the switch is closed, the light turns off. Which of the following statements is correct?

    (A) This can never happen, as switches allow current to flow through them when they are closed and the light needs current in order to turn on.
    (B) This implies that the switch is in series, either just before or just after the light.
    (C) This implies that the switch is in parallel with the light with no resistance along its parallel path.
    (D) This is how all switches normally operate: "open" means "on" and "closed" means "off."

25. Lenz's law is an electrical restatement of which conservation law?

    (A) linear momentum
    (B) angular momentum
    (C) charge
    (D) energy

26. A single electron traveling 2 cm away from a wire of current (2 amps) along a parallel trajectory to the wire's current will experience a force

    (A) of 0 N since the wire has no net charge.
    (B) directed radially away from the wire.
    (C) directed radially toward the wire.
    (D) parallel to the wire.

27. An electron is moving perpendicularly through a pair of crossed electric and magnetic fields (with magnitudes E and B, respectively) with constant velocity. The fields are perpendicular to each other. Which expression represents the speed of the electron?

    (A) EB
    (B) $2EB$
    (C) E/B
    (D) B/E

28. Refer to the figure below.

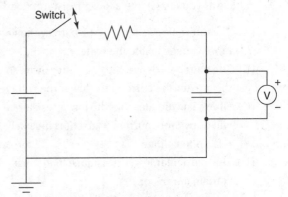

The switch is closed when the capacitor is uncharged. A long time later, the circuit reaches steady-state conditions. Which of the following choices correctly indicates how the current, I, in the switch and the magnitude of the potential difference, V, across the capacitor change during the charging time?

	I	V
(A)	increases	increases
(B)	increases	decreases
(C)	decreases	decreases
(D)	decreases	increases

29. An object is placed 15 centimeters in front of a concave spherical mirror whose radius of curvature is 12 centimeters. Which statement is correct about the image formed?

    (A) The image will be larger than the object and virtual.
    (B) The image will be smaller than the object and virtual.
    (C) The image will be larger than the object and real.
    (D) The image will be smaller than the object and real.

30. The isotope decays $^{13}_{7}N$ by emitting a positron. Which of the following isotopes is the direct decay product?

    (A) $^{13}_{6}C$
    (B) $^{12}_{6}C$
    (C) $^{12}_{7}N$
    (D) $^{13}_{8}O$

31. A 0.100 kg piece of an unknown metal is initially at a temperature of 392 K. The metal is then placed into 0.150 kg of water at a temperature of 295 K in an insulated container. The water and the metal come to equilibrium at a temperature of 302 K. The specific heat of water is 4,184 J/(kg · K). What is the specific heat of the unknown metal?

    (A) 325 J/(kg · K)
    (B) 488 J/(kg · K)
    (C) 650 J/(kg · K)
    (D) 6,276 J/(kg · K)

32. If a permanent magnet is cut in half halfway between its north and south poles, you will then have

    (A) a north pole piece strongly attracted to the south pole piece.
    (B) two pieces of magnetic material with no overall poles.
    (C) two smaller magnets, each with its own north and south poles.
    (D) a north pole piece slightly positive and a south pole piece slightly negative.

33. A solid conducting sphere is carrying a certain amount of overall positive charge. The electric potential both within and without will

	Inside Sphere	Outside Sphere
(A)	Gradually increase from center to exterior of sphere	Decrease as $1/r$
(B)	Gradually increase from center to exterior of sphere	Decrease as $1/r^2$
(C)	Decrease as $1/r$	Decrease as $1/r$
(D)	Be constant	Decrease as $1/r$

34. The voltage between the plates of a fully charged parallel plate capacitor will

    (A) be constant throughout.
    (B) fall linearly as you move from the positive plate to the negative plate.
    (C) fall quadratically as you move from the positive plate to the negative plate.
    (D) fall exponentially as you move from the positive plate to the negative plate.

35. What does the electric field resemble if the field is a far distance outside of a fully charged 5 F capacitor that is charged by a 12 V battery?

    (A) The electric field will resemble that of a 60 C point charge.
    (B) The electric field will be uniform and directed toward the capacitor.
    (C) The electric field will be uniform and directed away from the capacitor.
    (D) There will be no measurable electric field due to the capacitor.

36. An electron is observed to be moving in a clockwise circular path when viewed from above. Which statement best describes the magnetic field causing this?

    (A) The magnetic field is directed upward.
    (B) The magnetic field is directed downward.
    (C) The magnetic field is directed radially in the place of the circular path.
    (D) No magnetic field is present.

GO ON TO THE NEXT PAGE

PRACTICE TEST 2

PRACTICE TEST 2

37. A perpetual motion machine is an object with moving internal parts that never stop moving. Which concept best explains why an isolated perpetual motion machine is not possible?

    (A) conservation of energy
    (B) thermal conductivity
    (C) increase of entropy
    (D) conservation of momentum

38. One mole of helium is compared with one mole of nitrogen gas. Both gases are at the same temperature and pressure. Which statement is true?

    (A) The helium gas takes up less volume than the nitrogen gas because helium is a lighter molecule.
    (B) The helium gas molecules are moving faster on average than the nitrogen molecules because helium is a lighter molecule.
    (C) The speed of sound through the nitrogen gas will be faster since the nitrogen molecule is heavier.
    (D) The force delivered per area by the bouncing molecules of helium will be smaller than that from the nitrogen molecules since helium is a lighter molecule.

39. The fact that two oncoming headlights of a car cannot be resolved at large distances (i.e., the two light sources are seen as one) is best described by

    (A) wave interference.
    (B) refraction.
    (C) dispersion.
    (D) diffraction.

40. The Sun's light strikes the ground and heats it, which in turn heats the air in contact with the ground. This warm air rises and allows cooler air to rush in. What is the sequence of thermal energy transfer events?

    (A) convection → conduction → convection
    (B) radiation → conduction → convection
    (C) radiation → convection → conduction
    (D) conduction → conduction → convection

STOP   If there is still time remaining, you may review your answers.

Section II: Free-Response

TIME: 100 MINUTES

DIRECTIONS: Answer all parts of each of the following free-response questions using complete sentences. You may use a calculator and make use of the formula sheet provided in the Appendices.

---

1. (10 points; suggested time: 20–25 minutes)

A ray of light is incident on an unknown transparent substance from the air. The angle of incidence is θ_1, and the angle of refraction is θ_2. Note that $\theta_1 > \theta_2$.

(a) Explain why these measurements indicate that a change in speed must be happening as the light changes medium. Qualitatively **justify** whether the speed of light in the unknown transparent substance is higher or lower than that of air.

(b) **Derive** an expression for:

   i. the absolute index of refraction for this substance

   ii. the velocity of light in this substance

(c) The substance is now submerged in glycerol ($n = 1.47$). If $\theta_1 = 40°$ and $\theta_2 = 22°$, **calculate** the critical angle of incidence for light going from the unknown substance into glycerol. Explain why there is no critical angle when going the other way, from glycerol into the substance.

(d) The substance is now shaped into a convex lens. How does the focal length of this lens compare with the focal length of a similar-shaped lens made out of crown glass ($n = 1.52$)? **Justify** your answer.

GO ON TO THE NEXT PAGE

PRACTICE TEST 2

2. (12 points; suggested time: 25–30 minutes)

Use the electric field drawing below to answer the questions. *A*, *B*, and *C* are physical sources of charge. Points *e*, *d*, and *f* are points within the field.

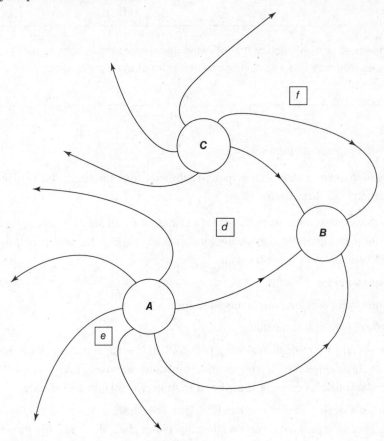

(a) **Describe** the relative amount of and type of charge (positive or negative) at each of the locations *A*, *B*, and *C*. Explain your answers.

(b) **Rank** the relative strength of the net electric field at each of points *e*, *d*, and *f*. Explain your answer.

(c) Using the expression "*A* + *B* + *C*" as the net charge, write an expression for the electric field at a great distance away (well outside of the picture given). **Sketch** out what the field would look like far away.

(d) On the original sketch above, **draw** a single, complete equipotential surface that runs through point *f*.

(e) If a proton were placed at point *f*, in what direction (if any) would it experience a force? **Determine** whether its change in electric potential energy would be negative or positive as it moved in response to this electric force. Explain your reasoning.

(f) **Indicate** whether both, one, or neither of these statements are true about the empty space at point *f*:

  ____ There is electric potential at point *f* but no electric potential energy.

  ____ There is an electric field at point *f* but no electric force.

  Justify your answer using physics principles.

3. (10 points; suggested time: 25–30 minutes)

A group of students is given 2 resistors (100 ohm and 200 ohm), a single capacitor (15 mF), two switches, 2 voltmeters, 2 ammeters, a 12-volt battery, and many alligator clip–style wires that can be used to connect these elements. The students are instructed to use these elements to determine the effect of adding a capacitor in parallel to an existing simple circuit.

(a) **Describe** how these elements could be set up to monitor the voltage across both resistors and the current through them. The capacitor, along with one of the resistors, should be controlled by a switch such that they can (or cannot) be part of the circuit as the switch is opened or closed. The second resistor should be in parallel with the capacitor-resistor combination. You should include a labeled circuit diagram of your setup to help in your description.

(b) **Indicate** what procedure to follow with your circuit design in order to determine the voltage and current of both resistors with and without the capacitor in the circuit. Include enough detail so that another student could carry out your procedure.

(c)   i. Based on the given values and assuming ideal conditions, **predict** the results of your procedure in part (b).

   ii. **Estimate** numerical predictions for steady-state behaviors.

   iii. **Sketch** a graphical solution for both current over time and voltage over time on the graph below.

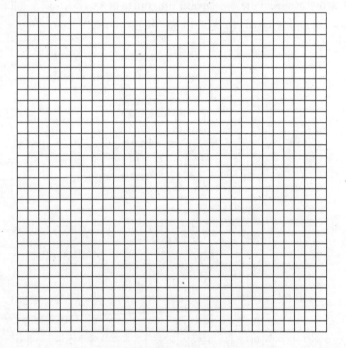

(d) **Describe** what effect the internal resistance of the battery might have on the students' actual results in part (c). Additionally, explain and describe the effects that any resistance within the ammeters themselves might have on the students' results.

(e) **Describe** qualitatively what would happen to the reading on both voltmeters and ammeters if, after the capacitor is fully charged, the battery is suddenly removed from the circuit, leaving an open circuit in its place.

GO ON TO THE NEXT PAGE

4. (8 points; suggested time: 15–20 minutes)

A student is investigating an unknown radioactive source with a Geiger counter. The Geiger counter clicks when it detects ionizing radiation, but it does not distinguish between alpha, beta, and gamma particles. The number of clicks indicates the number of ionizing events happening in the Geiger counter device due to the radiation it receives.

The student puts a piece of paper between the radioactive source and the Geiger counter and does not notice a change in the clicking rate. However, the addition of one sheet of aluminum foil does decrease the rate. When several sheets of aluminum foil are layered in a thickness of about a centimeter, there are no clicks at all.

(a) What type of radiation is being emitted? **Justify** your answer by describing alpha, beta, and gamma particles and how they interact with matter.

(b) **Indicate** what is happening to the atomic number and mass number within each atom that emits this radiation. Is it possible to predict which radioactive atoms within the sample will be emitting the radiation next?

(c) Is the binding energy per nucleon within the radioactive atoms changing as the radiation is being emitted? If so, is the binding energy per nucleon going up or down? **Justify** your answer with a qualitative explanation.

(d) If the half-life of the radioactive material being investigated is 12 hours, **sketch** a graph of the number of radioactive atoms in the source over the course of 3 days. Be sure to indicate (on the y-axis) the number of atoms left in terms of the starting amount, N_0, at $t = 0$.

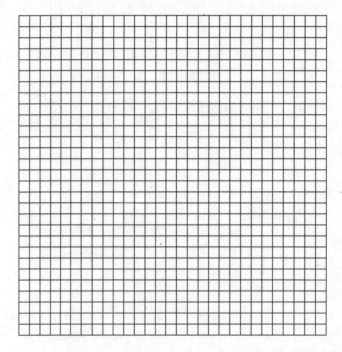

Answer Explanations

Section I

1. **(A)** Increasing the intensity will increase the number of photons per second but not change any aspect of the individual absorption event.

2. **(B)** Iron, with the highest binding energy per nucleon of any element listed, will not release any energy via fusion or fission.

3. **(C)** Compton's 1923 paper described his experiment in which momentum is conserved in a photon-electron collision.

4. **(D)** Both the mass number (top numbers) and the atomic number (bottom numbers) must add up to the same total before and after the reaction.

5. **(A)** The fusion of hydrogen into helium releases energy. So the helium nucleus must be lower in energy (more stable) than the previous nuclei.

6. **(C)** Rays going from slow to fast mediums refract away from the normal up until the critical angle, at which point no more refraction occurs.

7. **(B)** PV has units of joules in the metric system. Since $E = mc^2$, mc^2 must also have units of energy.

8. **(D)** Heat capacity is related to energy absorption and temperature change, not to the transfer of thermal energy (heat).

9. **(D)**
$$C = Q/\Delta V$$
$$Q = C\Delta V = (10 \times 10^{-6})(6) = 6 \times 10^{-5}\,\text{C}$$

10. **(A)** The charging is exponential and approaches the fixed value of the previous problem.

11. **(A)** $\frac{1}{2}C\Delta V^2 = \frac{1}{2}(10 \times 10^{-6})(6)^2 = 1.8 \times 10^{-4}\,\text{J}$

12. **(B)** Two rays are reflected: one from the top of the quarter-wavelength thickness and one from the bottom. The ray from the bottom will meet the top-layer reflection exactly one-half wavelength out of phase: destructive interference.

13. **(D)** Maximum magnitude will happen every half a wavelength (peak to trough). The speed of electromagnetic waves is 3×10^8 m/s. So the time for half a wave to pass is $3/(3 \times 10^8) = 10^{-8}$ s.

14. **(A)** Although an ideal battery will always maintain its terminal voltage, even when more current is drawn by a lower resistance load, real batteries do have internal resistance, which causes the voltage across their terminals to be lower when current is drawn. As more current is drawn, there is a larger voltage drop across the internal resistance, leaving less voltage across the terminals.

15. **(C)** $E = hf = hc/\lambda =$
$(6.63 \times 10^{-34})(3 \times 10^8)/(5.4 \times 10^{-7}) =$
3.68×10^{-19} J

16. **(A)**
$$d \sin \theta = m\lambda$$
$$dx/L = m\lambda$$
$$x = m\lambda L/d = (1)(5.4 \times 10^{-7})(2)/(1.0 \times 10^{-3}) = 1.08 \times 10^{-3}\,\text{m}$$

17. **(D)** The diffracted light from each slit spreads out and interferes with the light from the other slit to produce the pattern.

18. **(A)** The answer explanation for Question 16 shows that $x = m\lambda L/d$. If L increases, then x increases.

19. **(B)** The answer explanation for Question 16 shows that $x = m\lambda L/d$. If λ decreases, then x decreases.

20. **(C)** Changes in wavelength are proportional to changes in wave speed:
$$v_2 = v_1/3$$
$$n_2/n_1 = (c/v_1)/(c/v_2) = v_2/v_1 = 1/3$$

21. **(B)** Use Snell's law, and set the refracted angle to 90°:
$$2.0 \sin \theta_c = 1.3 \sin 90°$$
$$\theta_c = \sin^{-1}(1.3/2) = 40.5° \approx 41°$$

22. **(C)** Energy $= qV = (2e)(400\,\text{V}) = 800$ eV

23. **(B)** Objects become positive by losing electrons.

24. **(C)** An open switch is one that does not allow current through. Most switches that control lights are in series with the light so that when they are open, the light does not receive current and does not turn on. In this case, however, the opposite is happening; therefore, the closed switch must be diverting all current away from the bulb. When the switch is open, the alternate, no resistance, parallel path is no longer available, obliging the current to travel through the bulb.

25. **(D)** Conservation of energy is the underlying principle. By opposing the change in flux, Lenz's law is trying to keep the field energy constant.

26. **(B)** The wire sets up a circular magnetic field. The electron is traveling perpendicularly to the magnetic field. The right-hand rule dictates that the magnetic field lines are counterclockwise when looking directly at the oncoming current. Since the electron is negatively charged, the right-hand rule for forces results in an outward force:

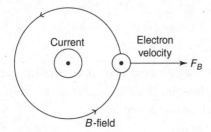

27. **(C)** Since the velocity is constant, the magnetic force must be canceling the electric force:

$$qE = qvB$$
$$v = E/B$$

28. **(D)** Once the capacitor is in the circuit, it will accumulate charge. As it charges, it will build up a potential difference, V. As the potential difference across the capacitor increases, the current in the circuit will decrease until the capacitor's voltage difference is the same as the battery's voltage difference (at which point the current will be zero).

29. **(D)** The object is located beyond the center of curvature. Therefore, the image will be real, inverted, and diminished.

30. **(A)** Positron emission reduces the atomic number by 1 but leaves the mass number unchanged.

31. **(B)** Heat flowing out of the hot metal must be equal to the heat gained by the water. Let x be the specific heat of the metal:

Heat out of metal = heat into water
$$mc\Delta T = mc\Delta T$$
$$(0.1)(x)(392 - 302) = (0.15)(4{,}184)(302 - 295)$$
$$0.1x\,(392 - 302) = 4{,}393.2$$
$$x = 488 \text{ J/(kg} \cdot \text{K)}$$

32. **(C)** At the break point, a new north pole and a new south pole will be found such that each smaller piece will have its own north and south poles. The two poles of a magnet can never be isolated.

33. **(D)** Since the sphere is a conductor, all excess charge will reside on the surface. The interior will be one constant value (determined by the value on the surface of the sphere) since there is no electric field. On the exterior of the sphere, the fields will be the same as if the excess charge were a point charge at the very center, so the force field will fall as $1/r^2$ and the energy field will fall as $1/r$.

34. **(B)** The constant electric field implies a linear fall in voltage as you move from the positive plate to the negative plate.

35. **(D)** The net charge on a capacitor is zero as it carries an equal amount of opposite charge. From far away, no dipole characteristic will emerge, and no field will be detected.

36. **(B)** There must be an inwardly directed force to keep the electron circulating. The electron moving clockwise represents a counterclockwise current. The right-hand rule requires a downward magnetic field to produce an inward force.

37. **(C)** Since the entropy of an isolated system must increase, some heat must be generated by the operation of the machine. This heat comes at the loss of available energy to make the machine work. Over time, the machine will run out of energy due to the increasing entropy.

38. **(B)** Since both gases are at the same temperature, their molecules must have the same average kinetic energy. Therefore, the less massive helium molecules must be moving faster in order to produce the same $\frac{1}{2}mv^2$.

39. **(D)** As the light leaves the headlights, they begin to diffract. From far enough away, each beam of light has diffracted such that their central maximums overlap and thus can no longer be distinguished.

40. **(B)** The Sun's light being absorbed by the ground is radiation. The ground heating the air in contact with it is conduction. Air movement caused by heat is convection.

PRACTICE TEST 2

Section II

1. (a) The change in angle as the ray of light changes medium is due to a change in speed. Just as a rolling cylinder will pivot if one end of it comes into contact with a rougher surface, so too will the ray of light pivot as it changes speed when crossing from one medium into another. Since the light is pivoting toward the normal, the new medium must be a slower one than air.

   [1 point for providing an appropriate explanation, and 1 point for providing clear justification]

   (b)   i. Starting with Snell's law:

$$n_1 \sin \theta_1 = n_2 \sin \theta_2$$

   Since air has the index of refraction $n_1 = 1.00$, we can solve for n_2:

$$n_2 = \frac{\sin \theta_1}{\sin \theta_2}$$

   [1 point]

   ii. Using the definition for the index of refraction:

$$n = \frac{c}{v}$$

   Substitute into the expression from part (b)(i):

$$\frac{c}{v} = \frac{\sin \theta_1}{\sin \theta_2}$$

   Solve for v:

$$v = \frac{c \sin \theta_2}{\sin \theta_1}$$

   [1 point]

   (c) For the critical angle, we know that the angle of refraction will be 90 degrees:

$$\sin \theta_c = \frac{n_1}{n_2}$$

   Next, we use the given angles to determine the index of refraction of the original medium:

$$n_2 = \frac{\sin \theta_1}{\sin \theta_2} = \frac{\sin 40°}{\sin 22°} = 1.72$$

   [1 point]

   Substitute in the values for n:

$$\sin \theta_c = \frac{1.47}{1.72} = 0.8547$$

$$\theta_c = 59°$$

   [1 point]

There is no critical angle when reversing the direction because total internal reflection can happen only when going from a slower medium to a faster one (i.e., higher n to lower n). This is so because going from a slower medium to a faster medium refracts the light away from the normal, toward the surface. This direction of refraction creates a limiting case (called the critical angle) that refracts light all the way to the surface itself (90-degree refraction). Angles beyond this don't work; no refraction occurs. When going from a faster medium to a slower medium, the light ray refracts toward the normal. So there is always a refracted solution and, hence, no critical angle. [2 points]

(d) The focal length of the lens made from the unknown substance will be shorter than the focal length of the lens made from crown glass. Since the index of refraction of the unknown substance is greater than that of crown glass, light will refract more. Thus, the converging point for the focal length will be closer to the lens.

[1 point for noting that the focal length of the lens made from the unknown substance will be shortened, and 1 point for providing clear justification]

2. (a) The number of field lines indicates the amount of charge. The direction of field lines indicates the type of charge (away = positive, toward = negative). Thus:

amount of charge A (positive) > amount of charge C (positive) > amount of charge B (negative)

[1 point for the proper ordering of the magnitudes, and 1 point for including proper signs on the three charges]

(b) The strength of the field is proportional to the density of the field lines:

field at e > field at f > field at d

[1 point]

(c) From very far away, the details of the actual charge distribution will not matter.

The field will look like that of a positive point charge with the seven unconnected field lines coming out uniformly. Below is an expression that represents this distant viewpoint as well as a sketch of what the field would look like far away.

$$E = k(A + B + C)/r^2$$

[1 point for a correct expression for the electric field]

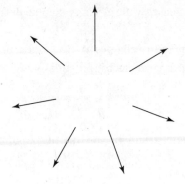

[2 points for a sketch of outward radial field lines]

(d) Equipotential lines should intersect electric field lines perpendicularly, do not have direction, and are continuous.

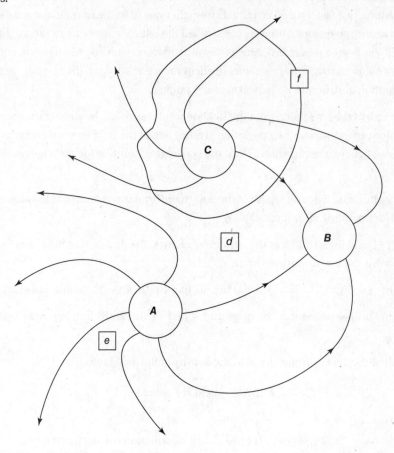

[2 points]

(e) A proton would follow the field direction at its location, so it experiences a force up and to the right at point *f*. [1 point]

Positive charges go from high voltage to low voltage:

$$\Delta V < 0, \text{ so } q\Delta V < 0$$

Electric potential energy decreases while electric potential decreases. [1 point]

(f) Both statements are true. The source charges (*A*, *B*, *C*) create fields (both electric fields (*E*) and voltage fields (*V*)), but for an electric force or electric potential energy, there must be an additional charge (*q*) at location *f*:

$$F = qE \qquad U_e = qV$$

[1 point for indicating that both statements are true, and 1 point for providing clear justification]

3. (a) Ammeters must be in series with their resistors. Voltmeters must be in parallel with their resistors. The capacitor must be in series with a switch and in parallel with one of the resistors in order to have a complete circuit when the switch is open. In order to not create a temporary short circuit when the switch is closed, a second resistor must be placed in series with the capacitor. [1 point]

One possible circuit diagram would look like this:

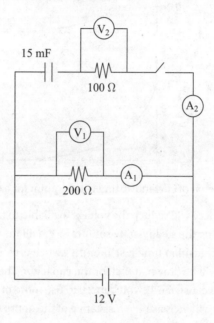

[1 point]

(b) After connecting the circuit above, leave the switch open. Then check the reading of the ammeter and voltmeter along the bottom pathway. (The other ammeter and voltmeter should each read zero.) Now, close the switch and wait for the readings to settle down on all four meters. This will not happen instantaneously. [1 point]

(c) i. With the switch open, A_2 and V_2 will read zero as they are not connected. What remains is a simple one resistor closed loop. After the switch is closed, current will flow through A_2 initially and V_2 will read a maximum voltage. Over time, V_2 and A_2 will decrease as the capacitor charges until they eventually both read zero when the capacitor is fully charged. [1 point]

    ii. With the switch open:

$$V_2 = 0$$

$$V_1 = 12\text{ V}$$

$$I_2 = 0$$

$$I_1 = 0.06\text{ A}$$

[1 point]

After the switch is closed, V_2 and I_2 will initially pop up to 12 V and 0.12 A, respectively. However, once the capacitor is fully charged, the steady-state solution with the switch closed will be the same as above (since the 12 volts will be across the capacitor!).

[1 point]

iii.

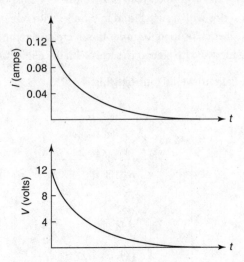

[1 point for a correct sketch of current vs. time, and 1 point for a correct sketch of voltage vs. time]

(d) Internal resistance of the battery will reduce the voltage available to the circuit as a function of the current drawn. Specifically, during the steady-state solutions, V_1 will measure less than 12 V and I_1 will be less than 0.06 A. During the transition time just after the switch is closed, the voltage and current will fall further as more current will be drawn at first for the capacitor. The greater current drawn during the charging of the capacitor will cause the 12-volt battery to use more of its voltage on its internal resistance. Ammeters having resistance will increase the resistance of the entire circuit, meaning a lower current drawn from the battery. This lower current will result in less voltage actually being used than the expected 12 V across the resistor. [1 point]

(e) There would still be a closed path for the capacitor to discharge through because both resistors would now be in series. The capacitor would quickly discharge. Both ammeters would have the same readings, which would quickly go to zero. V_1 would be twice that of V_2 while the capacitor discharges, but they, too, would quickly go to zero. [1 point]

4. (a) Beta particles are being emitted. [1 point]

Alpha particles are stopped by paper, and gamma rays would penetrate the aluminum foil easily. Alpha particles are so easily stopped because they are big and highly charged. Beta particles have a single quantum of charge and are quite small, so they are harder to stop. Gamma rays are high-frequency electromagnetic radiation and carry no charge. Thus, they are the most difficult to stop. [1 point]

(b) Upon emission of a beta particle, the atom will lose 4 units of mass and have its atomic number reduced by 2. [1 point]

It is not possible to predict which specific radioactive atom(s) will decay next. The process is random but defined by a certain probability. [1 point]

(c) The daughter nuclei will be more stable than before the beta particle was emitted. Therefore, its binding energy per nucleon will be higher. Another way to think about this is that the ejected electron is taking away energy in addition to its mass, leaving the daughter nuclei with a greater mass deficit per nucleon than before ($E = mc^2$).

[1 point for stating that the binding energy will be higher, and 1 point for providing clear justification]

(d)

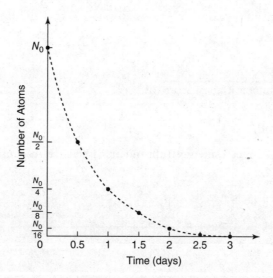

[1 point for proper labeling of the fractions of N_0 along the y-axis, and 1 point for approximating an exponential decrease over time]

Test Analysis

Section I: Multiple-Choice

Number correct (out of 40) = $\dfrac{}{\text{Multiple-Choice Score}}$

Section II: Free-Response

Partial credit is awarded for any correct responses within an individual free-response question.

Question 1 = $\dfrac{}{\text{(out of 10)}}$

Question 2 = $\dfrac{}{\text{(out of 12)}}$

Question 3 = $\dfrac{}{\text{(out of 10)}}$

Question 4 = $\dfrac{}{\text{(out of 8)}}$

Total (out of 40) = _____

Final Score

$$\dfrac{}{\text{Multiple-Choice Score}} + \dfrac{}{\text{Free-Response Score}} = \dfrac{}{\text{Total (out of 80)}}$$

Final Score Range*

Final Score Range	AP Score
60–80	5
44–59	4
32–43	3
20–31	2
0–19	1

*Note: Actual score ranges vary from year to year and are determined by the College Board each year. Thus, the ranges shown are approximate.

PRACTICE TEST 2

Appendices

Table of Information for AP Physics 2

Useful Constants	
Rest mass of the proton	$m_p = 1.67 \times 10^{-27}$ kg
Rest mass of the neutron	$m_n = 1.67 \times 10^{-27}$ kg
Rest mass of the electron	$m_e = 9.11 \times 10^{-31}$ kg
Magnitude of the electron charge	$e = 1.60 \times 10^{-19}$ C
Avogadro's number	$N_0 = 6.02 \times 10^{23}$ per mol
Universal gas constant	$R = 8.31$ J/(mol · K)
Boltzmann's constant	$k_B = 1.38 \times 10^{-23}$ J/K
Speed of light	$c = 3 \times 10^8$ m/s
Planck's constant	$h = 6.63 \times 10^{-34}$ J · s $= 4.14 \times 10^{-15}$ eV · s
1 electron volt	$1 \text{ eV} = 1.6 \times 10^{-19}$ J
Vacuum permittivity	$\varepsilon_0 = 8.85 \times 10^{-12}$ C^2/(N · m^2)
Coulomb's law constant	$k = (1/4)\pi\varepsilon_0 = 9 \times 10^9$ (N · m^2)/C^2
Vacuum permeability	$\mu_0 = 4\pi \times 10^{-7}$ Wb/(A · m)
Magnetic constant	$k' = \mu_0/4\pi = 10^{-7}$ Wb/(A · m)
Acceleration due to gravity at Earth's surface	$g = 9.8$ m/s^2
Universal gravitational constant	$G = 6.67 \times 10^{-11}$ m^3/(kg · s^2)
1 atmosphere of pressure	$1 \text{ atm} = 1.0 \times 10^5$ N/m$^2 = 1.0 \times 10^5$ Pa
Wien's constant	$b = 2.90 \times 10^{-3}$ m · K
Stefan-Boltzmann constant	$\sigma = 5.67 \times 10^{-8}$ W/(m^2 · K^4)
1 unified atomic mass unit	$1 \text{ u} = 1.66 \times 10^{-27}$ kg $= 931$ MeV/c^2

Unit Symbols			
Meter, m	Mole, mol	Watt, W	Farad, F
Kilogram, kg	Hertz, Hz	Coulomb, C	Tesla, T
Second, s	Newton, N	Volt, V	Degree Celsius, °C
Ampere, A	Pascal, Pa	Ohm, Ω	Electron volt, eV
Kelvin, K	Joule, J	Henry, H	

Prefixes		
Factor	**Prefix**	**Symbol**
10^{12}	Tera-	T
10^9	Giga-	G
10^6	Mega-	M
10^3	Kilo-	k
10^{-2}	Centi-	c
10^{-3}	Milli-	m
10^{-6}	Micro-	μ
10^{-9}	Nano-	n
10^{-12}	Pico-	p

Values of Trigonometric Functions for Common Angles							
θ	0°	30°	37°	45°	53°	60°	90°
$\sin\theta$	0	$\frac{1}{2}$	$\frac{3}{5}$	$\frac{\sqrt{2}}{2}$	$\frac{4}{5}$	$\frac{\sqrt{3}}{2}$	1
$\cos\theta$	1	$\frac{\sqrt{3}}{2}$	$\frac{4}{5}$	$\frac{\sqrt{2}}{2}$	$\frac{3}{5}$	$\frac{1}{2}$	0
$\tan\theta$	0	$\frac{\sqrt{3}}{3}$	$\frac{3}{4}$	1	$\frac{4}{3}$	$\sqrt{3}$	∞

Formula Sheet for AP Physics 2

Mechanics	Electricity and Magnetism

Mechanics

$\vec{v} = \vec{v}_0 + \vec{a}t$

$\vec{x} = \vec{x}_0 + \vec{v}_0 t + \frac{1}{2}\vec{a}t^2$

$v^2 = v_0{}^2 + 2a(x - x_0)$

$\sum \vec{F} = \vec{F}_{\text{net}} = m\vec{a}$

$F_{\text{fric}} \leq \mu N$

$a_c = \dfrac{v^2}{r}$

$\tau = rF\sin\theta$

$\vec{p} = m\vec{v}$

$\vec{J} = \vec{F}\Delta t = \Delta\vec{p}$

$K = \frac{1}{2}mv^2$

$\Delta U_g = mgh$

$W = F\Delta r\cos\theta$

$P_{\text{avg}} = \dfrac{\Delta W}{\Delta t}$

$\theta = \theta_0 + \omega_0 t + \frac{1}{2}\alpha t^2$

$\omega = \omega_0 + \alpha t$

$x = A\cos(\omega t) = A\cos(2\pi ft)$

$x_{\text{cm}} = \Sigma m_i x_i / \Sigma m_i$

$\sum \vec{\tau} = \vec{\tau}_{\text{net}} = I\vec{\alpha}$

$T = rF\sin\theta$

$\vec{L} = I\vec{\omega}$

$\Delta\vec{L} = \vec{\tau}\Delta t$

$K = \frac{1}{2}I\omega^2$

$P = Fv\cos\theta$

$\vec{F}_S = -k\vec{x}$

$U_s = \frac{1}{2}kx^2$

$T_S = 2\pi\sqrt{\dfrac{m}{k}}$

$T_p = 2\pi\sqrt{\dfrac{\ell}{g}}$

$T = \dfrac{1}{f}$

$F_G = -\dfrac{Gm_1 m_2}{r^2}$

$U_G = -\dfrac{Gm_1 m_2}{r}$

Electricity and Magnetism

$F = \dfrac{1}{4\pi\varepsilon_0}\dfrac{q_1 q_2}{r^2}$

$\vec{E} = \dfrac{\vec{F}_E}{q}$

$U_E = q\Delta v = \dfrac{1}{4\pi\varepsilon_0}\dfrac{q_1 q_2}{r^2}$

$E_{\text{avg}} = -\dfrac{\Delta V}{d}$

$V = \dfrac{1}{4\pi\varepsilon_0}\dfrac{q}{r}$

$C = \dfrac{Q}{\Delta V}$

$C = \dfrac{\varepsilon_0 A}{d}$

$U_c = \frac{1}{2}Q\Delta V = \frac{1}{2}C(\Delta V)^2$

$I_{\text{avg}} = \dfrac{\Delta Q}{\Delta t}$

$R = \dfrac{\rho\ell}{A}$

$E = \dfrac{q}{4\pi\varepsilon_0 r^2}$

$\Delta V = IR$

$P = I\Delta V$

$C_p = \sum_i C_i$

$\dfrac{1}{C_s} = \sum_i \dfrac{1}{C_i}$

$R_s = \sum_i R_i$

$\dfrac{1}{R_p} = \sum_i \dfrac{1}{R_i}$

$F_B = qvB\sin\theta$

$F_B = BI\ell\sin\theta$

$B = \dfrac{\mu_0 I}{2\pi r}$

$\phi_m = BA\cos\theta$

$\varepsilon_{\text{avg}} = -\dfrac{\Delta\phi_m}{\Delta t}$

$\varepsilon = B\ell v$

Mechanics symbols

a = acceleration
F = force
f = frequency
h = height
I = rotational inertia
J = impulse
K = kinetic energy
k = spring constant
L = angular momentum
ℓ = length
m = mass
N = normal force
P = power
p = momentum
r = radius or distance
T = period

t = time
U = potential energy
V = volume
v = velocity or speed
W = work done on a system
x = position
α = angular acceleration
θ = angle
τ = torque
ω = angular speed
μ = coefficient of friction

Electricity and Magnetism symbols

A = area
B = magnetic field
C = capacitance
d = distance
E = electric field
ε = emf
F = force
I = current
ℓ = length
P = power
Q = charge
q = point charge
R = resistance

r = distance
t = time
U = potential (stored) energy
V = electric potential
v = velocity or speed
ρ = resistivity
θ = angle
ϕ_m = magnetic flux

Fluid Mechanics and Thermal Physics

$$P = P_0 + \rho g h$$
$$F_{\text{buoy}} = \rho V g$$
$$A_1 v_1 = A_2 v_2$$
$$P + \rho g y + \frac{1}{2}\rho v^2 =$$
const.

$$P = \frac{F}{A}$$

$$PV = nRT = Nk_B T$$

$$\rho = m/V$$

$$\frac{Q}{\Delta t} = \frac{kA\Delta T}{L}$$

$$K = \frac{3}{2}k_B T$$
$$W = -P\Delta V$$
$$\Delta U = Q + W$$

A = area
F = force
h = depth
k = thermal conductivity
K = kinetic energy
L = thickness
n = number of moles
N = number of atoms
P = pressure
Q = energy
  transferred to a
  system by heating

T = temperature
t = time
U = internal energy
V = volume
v = velocity or speed
W = work done on
  a system
y = height
ρ = density

Modern Physics

$$E = hf$$
$$K_{\text{max}} = hf - \phi$$
$$E = mc^2$$
$$\lambda_{\text{max}} = \frac{b}{T}$$
$$P = A\sigma T^4$$
$$\Delta\lambda = \frac{h}{m_e c}(1 - \cos\theta)$$

$$E = mc^2$$
$$N = N_0 e^{-\lambda t}$$
$$\lambda = \frac{\ln 2}{t_{1/2}}$$
$$\lambda = \frac{h}{p}$$

E = energy
f = frequency
K = kinetic energy
m = mass

p = momentum
λ = wavelength
ϕ = work function

Waves, Sound, and Optics

$$v = f\lambda$$
$$n = \frac{c}{v}$$
$$n_1 \sin\theta_1 = n_2 \sin\theta_2$$
$$\frac{1}{s_i} + \frac{1}{s_0} = \frac{1}{f}$$
$$a\left(\frac{y_{\text{min}}}{L}\right) \approx m\lambda$$
$$\Delta D = d\sin\theta$$
$$d\left(\frac{y_{\text{max}}}{L}\right) \approx m\lambda$$

$$v_{\text{string}} = \sqrt{\frac{F_T}{m/\ell}}$$
$$T = \frac{1}{f}$$
$$x(t) = A\cos(\omega t) = A\cos(2\pi ft)$$
$$y(x) = A\cos\left(2\pi\frac{x}{\lambda}\right)$$
$$|f_{\text{beat}}| = |f_1 - f_2|$$
$$M = \frac{h_i}{h_0} = -\frac{s_i}{s_0}$$

d = separation
f = frequency or
  focal length
h = height
L = distance
M = magnification
m = an integer

n = index of refraction
s = position
v = speed
λ = wavelength
θ = angle

Geometry and Trigonometry

Rectangle
  $A = bh$
Triangle
  $A = \frac{1}{2}bh$
Circle
  $A = \pi r^2$
  $C = 2\pi r$
Rectangular Solid
  $V = \ell w h$
Cylinder
  $V = \pi r^2 \ell$
  $S = 2\pi r\ell + 2\pi r^2$
Sphere
  $V = \frac{4}{3}\pi r^3$
  $S = 4\pi r^2$

Right Triangle
  $a^2 + b^2 = c^2$
  $\sin\theta = \frac{a}{c}$
  $\cos\theta = \frac{b}{c}$
  $\tan\theta = \frac{a}{b}$

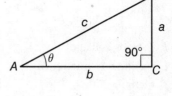

A = area
C = circumference
V = volume
S = surface area
b = base

h = height
ℓ = length
w = width
r = radius

Glossary

A

absolute index of refraction   For a transparent material, a number that represents the ratio of the speed of light in a vacuum to the speed of light in the material.

absolute temperature   A measure of the average kinetic energy of the molecules in an object; on the Kelvin scale, equal to the Celsius temperature of an object plus 273.

absolute zero   The theoretical lowest possible temperature, designated as 0 K.

absorption spectrum   A continuous spectrum crossed by dark lines representing the absorption of particular wavelengths of radiation by a cooler medium.

acceleration   A vector quantity representing the time rate of change of velocity.

action   A force applied to an object that leads to an equal but opposite reaction; the product of total energy and time in units of joules times seconds; the product of momentum and position, especially in the case of the Heisenberg uncertainty principle.

activity   The number of decays per second of a radioactive atom.

adiabatic   In thermodynamics, referring to the process that occurs when no heat is added or subtracted from the system.

alpha decay   The spontaneous emission of a helium nucleus from certain radioactive atoms.

alpha particle   A helium nucleus that is ejected from a radioactive atom.

alternating current   An electric current that changes its direction and magnitude according to a regular frequency.

ammeter   A device that, when placed in series, measures the current in an electric circuit; a galvanometer with a low-resistance coil placed across it.

ampere (A)   The SI unit of electric current, equal to 1 C/s.

amplitude   The maximum displacement of an oscillating particle, medium, or field relative to its rest position.

angle of incidence   The angle between a ray of light and the normal to a reflecting or transparent surface at the location where the ray intercepts the surface.

angle of reflection   The angle between a reflected light ray and the normal to a mirror or other reflecting surface at the location where the ray intercepts the surface.

angle of refraction   The angle between an emerging light ray in a transparent material and the normal to the surface at the location where the ray first enters the material.

angstrom (Å)   A unit of distance measurement equal to 1×10^{-10} m.

antinodal lines   A region of maximum displacement in a medium where waves are interacting with each other.

Archimedes' principle   A body wholly or partially immersed in a fluid will be buoyed up by a force equal to the weight of the fluid it displaces.

armature   The rotating coil of wire in an electric motor.

atmosphere (atm)   A unit of pressure, equal to 101 kPa at sea level.

atomic mass unit (u)   A unit of mass equal to one-twelfth the mass of a carbon-12 nucleus. Older texts use the abbreviation "amu."

atomic number   The number of protons in an atom's nucleus.

average speed   A scalar quantity equal to the ratio of the total distance to the total elapsed time.

Avogadro's number   The number of molecules in 1 mole of an ideal gas, equal to 6.02×10^{23}. Named for Italian chemist/physicist Amedeo Avogadro.

B

Balmer series   In a hydrogen atom, the visible spectral emission lines that correspond to electron transitions from higher excited states to lower level 2.

battery   A combination of two or more electric cells.

beats   The interference caused by two sets of sound waves with only a slight difference in frequency.

becquerel (Bq)   A unit of radioactive decay, equal to one decay event per second.

Bernoulli's principle   If the speed of a fluid particle increases as it travels along a streamline, the pressure of the fluid must decrease.

beta decay   The spontaneous ejection of an electron from the nuclei of certain radioactive atoms.

beta particle   An electron that is spontaneously ejected by a radioactive nucleus.

binding energy   The energy required to break apart an atomic nucleus; the energy equivalent of the mass defect.

Boltzmann's constant   In thermodynamics, a constant equal to the ratio of the universal gas constant, R, to Avogadro's number, N_A, equal to 1.38×10^{-23} J/K. Named for Austrian physicist Ludwig Boltzmann.

Boyle's law   At constant temperature, the pressure in an ideal gas varies inversely with the volume of the gas. Named for Anglo-Irish physicist/chemist Robert Boyle.

bright-line spectrum   The display of brightly colored lines on a screen or photograph indicating the discrete emission of radiation by a heated gas at low pressure.

C

capacitance   The ratio of the total charge to the potential difference in a capacitor.

capacitor   A pair of conducting plates, with either a vacuum or an insulator between them, used in an electric circuit to store current.

Carnot cycle   In thermodynamics, a sequence of four steps in an ideal gas confined in a cylinder with a movable piston (a Carnot engine). The cycle includes an isothermal expansion, an adiabatic expansion, an isothermal compression, and an adiabatic compression. Named for French physicist Sadi Carnot.

Cartesian coordinate system   A set of two or three mutually perpendicular reference lines, called axes and usually designated as x, y, and z, that are used to define the location of an object in a frame of reference; a coordinate system named for French scientist Rene Descartes.

cathode ray tube   An evacuated gas tube into which a beam of electrons is projected. Their energy produces an image on a fluorescent screen when deflected by external electric or magnetic fields.

Celsius temperature scale   A metric temperature scale in which, at sea level, water freezes at 0°C and boils at 100°C. Named for Swedish astronomer Anders Celsius.

center of curvature   A point that is equidistant from all other points on the surface of a spherical mirror; a point equal to twice the focal length of a spherical mirror.

center of mass   The weighted mean distribution point where all the mass of an object can be considered to be located; the point at which, if a single force is applied, translational motion will result.

centripetal acceleration   The acceleration of mass moving in a circular path directed radially inward toward the center of the circular path.

centripetal force   The deflecting force, directed radially inward toward a given point, that causes an object to follow a circular path.

chain reaction   In nuclear fission, the uncontrolled reaction of neutrons splitting uranium nuclei and creating more neutrons that continue the process on a self-sustained basis.

Charles's law   At constant pressure, the volume of an ideal gas varies directly with the absolute temperature of the gas. Named for J. A. C. Charles.

chromatic aberration   In optics, the defect in a converging lens that causes the dispersion of white light into a continuous spectrum, with the result that the lens refracts the colors to different focal points.

coefficient of friction   The ratio of the force of friction to the normal force when one surface is sliding (or attempting to slide) over another surface.

coherent   Referring to a set of waves that have the same wavelength, frequency, and phase.

component   One of two mutually perpendicular vectors that lie along the principal axes in a coordinate system and can be combined to form a given resultant vector.

concave lens   A diverging lens that causes parallel rays of light to emerge in such a way that they appear to diverge away from a focal point behind the lens.

concave mirror   A converging spherical mirror that causes parallel rays of light to converge to a focal point in front of the mirror.

concurrent forces   Two or more forces that act at the same point and at the same time.

conductor   A substance, usually metallic, that allows the relatively easy flow of electric charges.

conservation of energy   A principle of physics that states that the total energy of an isolated system remains the same during all interactions within the system.

conservation of mass-energy   A principle of physics that states that, in the conversion of mass into energy or energy into mass, the total mass-energy of the system remains the same.

conservation of momentum   A principle of physics that states that, in the absence of any external forces, the total momentum of an isolated system remains the same.

conservative force   A force such that any work done by this force can be recovered without any loss; a force whose work is independent of the path taken.

constructive interference   The additive result of two or more waves interacting with the same phase relationship as they move through a medium.

continuous spectrum   A continuous band of colors, consisting of red, orange, yellow, green, blue, and violet, formed by the dispersion or diffraction of white light.

control rod   A device, usually made of cadmium, that is inserted in a nuclear reactor to control the rate of fission.

converging lens   A lens that will cause parallel rays of light incident on its surface to refract and converge to a focal point; a convex lens.

converging mirror   A spherical mirror that will cause parallel rays of light incident on its surface to reflect and converge to a focal point; a concave mirror.

convex lens   See **converging lens**.

convex mirror   See **diverging mirror**.

coordinate system   A set of reference lines, not necessarily perpendicular, used to locate the position of an object within a frame of reference by applying the rules of analytic geometry.

core   The interior of a solenoidal electromagnet, usually made of a ferromagnetic material; the part of a nuclear reactor where the fission reaction occurs.

coulomb (C)   The SI unit of electric charge, defined as the amount of charge 1 A of current contains each second.

Coulomb's law   The electrostatic force between two point charges is directly proportional to the product of the charges and inversely proportional to the square of the distance separating them. Named for French physicist Charles-Augustin de Coulomb.

critical angle of incidence   The angle of incidence to a transparent substance such that the angle of refraction equals 90° relative to the normal drawn to the surface.

current   A scalar quantity that measures the amount of charge passing a given point in an electric circuit each second.

current length   A relative measure of the magnetic field strength produced by a length of wire carrying current, equal to the product of the current and the length.

cycle   One complete sequence of periodic events or oscillations.

D

damping   The continuous decrease in the amplitude of mechanical oscillations due to a dissipative force.

deflecting force   Any force that acts to change the direction of motion of an object.

derived unit   Any combination of fundamental physical units.

destructive interference   The result produced by the interaction of two or more waves with opposite phase relationships as they move through a medium.

deuterium   An isotope of hydrogen containing one proton and one neutron in its nucleus; heavy hydrogen (a component of heavy water).

dielectric   An electric insulator placed between the plates of a capacitor to alter its capacitance.

diffraction   The ability of waves to pass around obstacles or squeeze through small openings.

diffraction grating   A reflecting or transparent surface with many thousands of lines ruled on it, used to diffract light into a spectrum.

direct current   Electric current that is moving in one direction only around an electric circuit.

dispersion   The separation of light into its component colors or spectrum.

dispersive medium   Any medium that produces the dispersion of light; any medium in which the velocity of a wave depends on its frequency.

displacement   A vector quantity that determines the change in position of an object by measuring the straight-line distance and direction from the starting point to the ending point.

dissipative force   Any force, such as friction, that removes kinetic energy from a moving object; a nonconservative force.

distance   A scalar quantity that measures the total length of the path taken by a moving object.

diverging lens   A lens that causes parallel rays of light incident on its surface to refract and diverge away from a focal point on the other side of the lens; a concave lens.

diverging mirror   A spherical mirror that causes parallel rays of light incident on its surface to reflect and diverge away from a focal point on the other side of the mirror.

Doppler effect   The apparent change in the wavelength or frequency of a wave as the source of the wave moves relative to an observer. Named for Austrian physicist Christian Doppler.

dynamics   The branch of mechanics that studies the effects of forces on objects.

E

elastic collision   A collision between two objects in which there is a rebounding and no loss of kinetic energy occurs.

elastic potential energy   The energy stored in a spring when work is done to stretch or compress it.

electric cell   A chemical device for generating electricity.

electric circuit   A closed conducting loop consisting of a source of potential difference, conducting wires, and other devices that operate on electricity.

electric field   The region where an electric force is exerted on a charged object.

electric field intensity   A vector quantity that measures the ratio of the magnitude of the force to the magnitude of the charge on an object.

electrical ground   The passing of charges to or from Earth to establish a potential difference between two points.

electromagnet   A coil of wire, wrapped around a ferromagnetic core (usually made of iron), that generates a magnetic field when current is passed through it.

electromagnetic field   The field produced by an electromagnet or moving electric charges.

electromagnetic induction   The production of a potential difference in a conductor due to the relative motion between the conductor and an external magnetic field, or due to the change in an external magnetic flux near the conductor.

electromagnetic spectrum   The range of frequencies covering the discrete emission of energy from oscillating electromagnetic fields; included are radio waves, microwaves, infrared waves, visible light, ultraviolet light, X-rays, and gamma rays.

electromagnetic wave   A wave generated by the oscillation of electric charges producing interacting electric and magnetic fields that oscillate in space and travel at the speed of light in a vacuum.

electromotive force (emf)   The potential difference caused by the conversion of different forms of energy into electrical energy; the energy per unit charge.

electron   A negatively charged particle that orbits a nucleus in an atom; the fundamental carrier of negative electric charge.

electron capture   The process in which an orbiting electron is captured by a nucleus possessing too many neutrons with respect to protons; also called K-capture.

electron cloud   A theoretical probability distribution of electrons around the nucleus due to the Heisenberg uncertainty principle. The most probable location for an electron is in the densest regions of the cloud.

electron volt (eV)   A unit of energy related to the kinetic energy of a moving charge and equal to 1.6×10^{-19} J.

electroscope   A device for detecting the presence of static charges on an object.

elementary charge   The fundamental amount of charge of an electron.

emf   See **electromotive force**.

emission spectrum   The discrete set of colored lines representing the electromagnetic energy produced when atomic compounds are excited into emitting light because of heat, sparks, or atomic collisions.

energy   A scalar quantity representing the capacity to work.

energy level   One of several regions around a nucleus where electrons are considered to reside.

entropy   The degree of randomness or disorder in a thermodynamic system.

equilibrant   The force equal in magnitude and opposite in direction to the resultant of two or more forces that brings a system into equilibrium.

equilibrium   The balancing of all external forces acting on a mass; the result of a zero-vector sum of all forces acting on an object.

escape velocity   The velocity attained by an object such that, if coasting, the object would not be pulled back toward the planet from which it came.

excitation   The process by which an atom absorbs energy and causes its orbiting electrons to move to higher energy levels.

excited state   In an atom, the situation in which its orbiting electrons are residing in higher energy levels.

F

farad (F)   A unit of capacitance equal to 1 C/V.

Faraday's law of electromagnetic induction   The magnitude of the induced emf in a conductor is equal to the rate of change of the magnetic flux. Named for British chemist/physicist Michael Faraday.

ferromagnetic substances   A metal or a compound made of iron, cobalt, or nickel that produces very strong magnetic fields.

field   A region characterized by the presence of a force on a test body like a unit mass in a gravitational field or a unit charge in an electric field.

field intensity   A measure of the force exerted on a unit test body; the force per unit mass; the force per unit charge.

first law of thermodynamics   A statement of the conservation of energy as applied to thermodynamic systems: The change in energy of a system is equal to the change in the internal energy plus any work done by the system.

fission   The splitting of a uranium nucleus into two smaller, more stable nuclei by means of a slow-moving neutron, with the release of a large amount of energy.

flux   A measure, in webers, of the product of the perpendicular component of a magnetic field crossing an area and the magnitude of the area.

flux density   A measure, in webers per square meter, of the field intensity per unit area.

focal length   The distance along the principal axis from a lens or spherical mirror to the principal focus.

focus   The point of convergence of light rays caused by a converging mirror or lens; either of two fixed points in an ellipse that determines its shape.

force   A vector quantity that corresponds to any push or pull due to an interaction of matter that changes the motion of an object.

force constant   See **spring constant**.

forced vibration   A vibration caused by the application of an external force.

frame of reference   A point of view consisting of a coordinate system in which observations are made.

free-body diagram   A diagram that illustrates all of the forces acting on a mass at any given time.

frequency   The number of completed periodic cycles per second in an oscillation or wave motion.

friction   A force that opposes the motion of an object as it slides over another surface.

fuel rods   Rods packed with fissionable material that are inserted into the core of a nuclear reactor.

fundamental unit   An arbitrary scale of measurement assigned to certain physical quantities, such as length, time, mass, and charge, that are considered to be the basis for all other measurements. In the SI system, the fundamental units used in physics are the meter, kilogram, second, ampere, kelvin, and mole.

fusion   The combination of two or more light nuclei to produce a more stable, heavier nucleus, with the release of energy.

G

galvanometer   A device used to detect the presence of small electric currents when connected in series in a circuit.

gamma radiation   High-energy photons emitted by certain radioactive substances.

gravitation   The mutual force of attraction between two uncharged masses.

gravitational field strength   A measure of the gravitational force per unit mass in a gravitational field.

gravity   Another name for gravitation or the gravitational force; the tendency of objects to fall to Earth.

ground state   The lowest energy level of an atom.

H

heat   The energy observed due to particle collisions in matter; the energy produced when matter interacts with particles that are colliding randomly with each other, as in a gas.

hertz (Hz)   The SI unit of frequency, equal to 1 s^{-1}.

Hooke's law   The stress applied to an elastic material is directly proportional to the strain produced. Named for English scientist Robert Hooke.

I

ideal gas   A gas for which the assumptions of the kinetic theory are valid.

image   An optical reproduction of an object by means of a lens or a mirror.

impulse   A vector quantity equal to the product of the average force applied to a mass and the time interval in which the force acts; the area under a force-versus-time graph.

induced potential difference   A potential difference created in a conductor because of its motion relative to an external magnetic field.

induction coil   A transformer in which a variable potential difference is produced in a secondary coil when a direct current, applied to the primary, is turned on and off.

inelastic collision   A collision in which two masses interact and stick together, leading to an apparent loss of kinetic energy.

inertia   The property of matter that resists the action of applied force trying to change the motion of an object.

inertial frame of reference   A frame of reference in which the law of inertia holds; a frame of reference moving with constant velocity relative to Earth.

instantaneous velocity   The slope of a tangent line to a point in a displacement-versus-time graph.

insulator   A substance that is a poor conductor of electricity because of the absence of free electrons.

interference   The interaction of two or more waves, producing an enhanced or a diminished amplitude at the point of interaction; the superposition of one wave on another.

interference pattern   The pattern produced by the constructive and destructive interference of waves generated by two point sources.

isobaric   In thermodynamics, referring to a process in which the pressure of a gas remains the same.

isochoric   In thermodynamics, referring to a process in which the volume of a gas remains the same.

isolated system   A combination of two or more interacting objects that are not being acted upon by an external force.

isotope   An atom with the same number of protons as a particular element but a different number of neutrons.

J

joule (J)   The SI unit of work, equal to $1 N \cdot m$; the SI unit of mechanical energy, equal to $1 kg \cdot m^2/s^2$.

junction   The point in an electric circuit where a parallel connection branches off.

K

K-capture   See **electron capture.**

kelvin (K)   The SI unit of absolute temperature, defined in such a way that 0 K equals $-273°C$.

Kepler's first law   The orbital paths of all planets are elliptical. Named for German astronomer Johannes Kepler.

Kepler's second law   A line from the Sun to a planet sweeps out equal areas in equal time.

Kepler's third law   The ratio of the cube of the mean radius to the Sun to the square of the period is a constant for all planets orbiting the Sun.

kilogram (kg)   The SI unit of mass.

kilojoule (kJ)   A unit representing 1,000 J.

kilopascal (kPa)   A unit representing 1,000 Pa.

kinematics   In mechanics, the study of how objects move.

kinetic energy   The energy possessed by a mass because of its motion relative to a frame of reference.

kinetic friction   The friction induced by sliding one surface over another.

kinetic theory of gases   The theory that all matter consists of molecules that are in a constant state of motion.

Kirchhoff's Junction Rule   The algebraic sum of all currents at a junction in a circuit equals zero. Named for German physicist Gustav Kirchhoff.

Kirchhoff's Loop Rule   The algebraic sum of all potential drops around any closed loop in a circuit equals zero.

L

laser   A device for producing an intense coherent beam of monochromatic light; an acronym for *l*ight *a*mplification by the *s*timulated *e*mission of *r*adiation.

law of inertia   See **Newton's first law of motion.**

lens   A transparent substance, with one or two curved surfaces, used to direct rays of light by refraction.

Lenz's law   An induced current in a conductor is always in a direction such that its magnetic field opposes the magnetic field that induced it. Named for German physicist Emil Lenz.

line of force   An imaginary line drawn in a gravitational, electrical, or magnetic field that indicates the direction a test particle would take while experiencing a force in that field.

longitudinal wave   A wave in which the oscillating particles vibrate in a direction parallel to the direction of propagation.

M

magnet   An object that exerts a force on ferrous materials.

magnetic field   A region surrounding a magnet in which ferrous materials or charged particles experience a force.

magnetic field strength   The force on a unit current length in a magnetic field.

magnetic flux density   The total magnetic flux per unit area in a magnetic field.

magnetic pole   A region on a magnet where magnetic lines of force are most concentrated.

mass   The property of matter used to represent the inertia of an object; the ratio of the net force applied to an object and its subsequent acceleration as specified by Newton's second law of motion.

mass defect   The difference between the actual mass of a nucleus and the sum of the masses of the protons and neutrons it contains.

mass number   The total number of protons and neutrons in an atomic nucleus.

mass spectrometer   A device that uses magnetic fields to cause nuclear ions to assume circular trajectories and then determines the masses of the ions based on their charges, the radius of the path, and the external field strength.

mean radius   For a planet, the average distance to the Sun.

meter (m)   The SI unit of length.

moderator   A material, such as water or graphite, that is used to slow neutrons in a fission reaction.

mole (mol)   The SI unit for the amount of substance containing Avogadro's number of molecules.

moment arm distance   A line drawn from a pivot point.

momentum   A vector quantity equal to the product of the mass and the velocity of a moving object.

monochromatic light   Light consisting of only one frequency.

N

natural frequency   The frequency with which an elastic body will vibrate if disturbed.

net force   The resultant force acting on a mass.

neutron   A subatomic particle, residing in the nucleus, that has a mass comparable to that of a proton but is electrically neutral.

newton (N)   The SI unit of force, equal to $1 \text{ kg} \cdot \text{m/s}^2$.

Newton's first law of motion   An object at rest tends to remain at rest, and an object in motion tends to remain in motion at a constant velocity, unless acted upon by an external force; the law of inertia. Named for British mathematician/physicist Sir Isaac Newton.

Newton's second law of motion   The acceleration of a body is directly proportional to the applied net force; $\vec{F} = m\vec{a}$.

Newton's third law of motion   For every action, there is an equal, but opposite, reaction.

nodal line   A line of minimum displacement when two or more waves interfere.

nonconservative force   A force, such as friction, that decreases the amount of kinetic energy after work is done by or against the force.

normal   A line perpendicular to a surface.

normal force   A force that is directed perpendicularly to a surface when two objects are in contact.

nuclear force   The force between nucleons in a nucleus that opposes the Coulomb force repulsion of the protons; the strong force.

nucleon   Either a proton or a neutron as it exists in a nucleus.

O

ohm (Ω)   The SI unit of electrical resistance, equal to 1 V/A.

Ohm's law In a circuit at constant temperature, the ratio of the potential difference to the current is a constant (called the resistance). Named for German physicist Georg Ohm.

optical center The geometric center of a converging or diverging lens.

P

parallel circuit A circuit in which two or more devices are connected across the same potential difference and provide an alternative path for charge flow.

pascal (Pa) The SI unit of pressure, equal to 1 N/m^2.

Pascal's principle A change in the pressure applied to an enclosed fluid is transmitted undiminished to every portion of the fluid and to the walls of the containing vessel.

period The time, in seconds, to complete one cycle of repetitive oscillations or uniform circular motion.

permeability The property of matter that affects an external magnetic field in which the material is placed, causing it to align free electrons within the substance and hence magnetizing it; a measure of a material's ability to become magnetized.

phase The relative position of a point on a wave with respect to another point on the same wave.

photoelectric effect The process by which surface electrons in a metal are freed through the incidence of electromagnetic radiation above a certain minimum frequency.

photon A single packet of electromagnetic energy.

Planck's constant A universal constant representing the ratio of the energy of a photon to its frequency; the fundamental quantum of action, having a value of $6.63 \times 10^{-34} \text{ kg} \cdot \text{m}^2/\text{s}$. Named for German physicist Max Planck.

polar coordinate system A coordinate system in which the location of a point is determined by a vector from the origin of a Cartesian coordinate system and the acute angle it makes with the positive horizontal axis.

polarization The process by which the vibrations of a transverse wave are selected to lie in a preferred plane perpendicular to the direction of propagation.

polarized light Light that has been polarized by passing it through a suitable filter.

positron A subatomic particle having the same mass as an electron but a positive electrical charge.

potential difference The difference in work per unit charge between any two points in an electric field.

potential energy The energy in a system resulting from the relative position of two objects interacting via a conservative force.

power The ratio of the work done to the time needed to complete the work.

pressure The force per unit area.

primary coil In a transformer, the coil connected to an alternating potential difference source.

principal axis An imaginary line passing through the center of curvature of a spherical mirror or the optical center of a lens.

prism A transparent triangular shape that disperses white light into a continuous spectrum.

pulse A single vibratory disturbance in an elastic medium.

Q

quantum A discrete packet of electromagnetic energy; a photon.

quantum theory The theory that light consists of discrete packets of energy that are absorbed or emitted in units.

R

radioactive decay The spontaneous disintegration of a nucleus due to its instability.

radius of curvature The distance from the center of curvature to the surface of a spherical mirror.

ray A straight line used to indicate the direction of travel of a light wave.

real image An image formed by a converging mirror or lens that can be focused onto a screen.

refraction   The bending of a light ray when it passes obliquely from one transparent substance to another in which its velocity is different.

resistance   The ratio of the potential difference across a conductor to the current in the conductor.

resistivity   A property of matter that measures the ability of that substance to act as a resistor.

resolution of forces   The process by which a given force is decomposed into a pair of perpendicular forces.

resonance   The production of sympathetic vibrations in a body, at its natural vibrating frequency, caused by another vibrating body.

rest mass   The mass of an object when the object is not moving.

rotational equilibrium   The situation when the vector sum of all torques acting on a rotating mass equals zero.

rotational inertia   The ability of a substance to resist the action of a torque; moment of inertia.

S

scalar   A physical quantity, such as mass or speed, that is characterized by magnitude only.

second (s)   The SI unit of time.

second law of thermodynamics   Heat flows only from a hot body to a cold one; the entropy of the universe is always increasing.

secondary coil   In a transformer, a coil in which an alternating potential difference is induced.

series circuit   A circuit in which electrical devices are connected sequentially in a single conducting loop, allowing only one path for charge flow.

sliding friction   A force that resists the sliding motion of one surface over another; kinetic friction.

Snell's law   For light passing obliquely from one transparent substance to another, the ratio of the sine of the angle of incidence to the sine of the angle of refraction is equal to the relative index of refraction for the two media. Named for Willebrord Snell.

solenoid   A coil of wire used for electromagnetic induction.

specific heat   The amount of energy, in kilojoules, needed to change the temperature of 1 kg of a substance by 1°C.

speed   A scalar quantity measuring the time rate of change of distance.

spherical aberration   In a converging spherical mirror or a converging lens, the inability to properly focus parallel rays of light because of the shape of the mirror or lens.

spring constant   The ratio of the applied force and resultant displacement of a spring.

standard pressure   At sea level, 1 atm, which is equal to 101.3 kPa.

standing wave   A stationary wave pattern formed in a medium when two sets of waves with equal wavelength and amplitude pass through each other, usually after a reflection.

static electricity   Stationary electric charges.

static friction   The force that prevents one object from sliding over another.

statics   The study of the forces acting on an object that is at rest relative to a frame of reference.

superposition   The ability of waves to pass through each other, interfere, and then continue on their way unimpeded.

T

temperature   The relative measure of warmth or cold relative to a standard; see also **absolute temperature**.

tesla (T)   The SI unit of magnetic field strength, equal to 1 W/m^2.

test charge   A small, positively charged mass used to detect the presence of a force in an electric field.

thermionic emission   The emission of electrons from a hot filament.

thermometer   An instrument that makes a quantitative measurement of temperature based on an accepted scale.

third law of thermodynamics   The temperature of any system can never be reduced to absolute zero.

threshold frequency The minimum frequency of electromagnetic radiation necessary to induce the photoelectric effect in a metal.

torque The application, from a pivot, of a force at right angles to a designated line that tends to produce circular motion; the product of a force and a perpendicular moment arm distance.

total internal reflection The process by which light is incident on a medium in which its velocity would increase at an angle greater than the critical angle of incidence.

total mechanical energy In a mechanical system, the sum of the kinetic and potential energies.

transmutation The process of changing one atomic nucleus into another by radioactive decay or nuclear bombardment.

transverse wave A wave in which the vibrations of the medium or field are at right angles to the direction of propagation.

U

uniform circular motion Motion around a circle at a constant speed.

uniform motion Motion at a constant speed in a straight line.

unit An arbitrary scale assigned to a physical quantity for measurement and comparison.

universal law of gravitation The gravitational force between any two masses is directly proportional to the product of their masses, and inversely proportional to the square of the distance between them.

V

vector A physical quantity that is characterized by both magnitude and direction; a directed arrow drawn in a Cartesian coordinate system used to represent a quantity such as force, velocity, or displacement.

velocity A vector quantity representing the time rate of change of displacement.

virtual focus The point at which the rays from a diverging lens would meet if they were traced back with straight lines.

virtual image An image, formed by a mirror or lens, that cannot be focused onto a screen.

visible light The portion of the electromagnetic spectrum that can be detected by the human eye.

volt (V) The SI unit of potential difference, equal to 1 J/C.

voltmeter A device used to measure the potential difference between two points in a circuit when connected in parallel; a galvanometer with a high-resistance coil placed in series with it.

volts per meter (V/m) The SI unit of electric field intensity, equal to 1 N/C.

W

watt (W) The SI unit of power, equal to 1 J/s.

wave A series of periodic disturbances in an elastic medium or field.

wavelength The distance between any two successive points in phase on a wave.

weber (Wb) The SI unit of magnetic flux, equal to $1 \text{ T} \times 1 \text{ m}^2$.

weight The force of gravity exerted on a mass at the surface of a planet.

work A scalar measure of the relative amount of change in mechanical energy; the product of the magnitude of the displacement of an object and the component of applied force in the same direction as the displacement; the area under a graph of force versus displacement.

work function The minimum amount of energy needed to free an electron from the surface of a metal using the photoelectric effect.

Index